Analysing Systems

BCS Practitioner Series

Series editor: Ray Welland

BELINA ET AL   SDL: with applications from protocol specification
BRAEK/HAUGEN   Engineering real time systems
BRINKWORTH   Software quality management: a pro-active approach
CRITCHLEY/BATTY   Open systems – the reality
FOLKES/STUBENVOLL   Accelerated systems development
GIBSON   Managing computer projects: avoiding the pitfalls
HIPPERSON   Practical systems analysis: for users, managers and analysts
HORROCKS/MOSS   Practical data administration
LEYLAND   Electronic data interchange
LOW   Writing user documentation: a practical guide for those who want to be read
MONK ET AL   Improving your human–computer interface: a practical technique
O'CONNELL   How to run successful projects
THE RAISE LANGUAGE GROUP   The RAISE specification language
RICE   VMS systems management
TANSLEY/HAYBALL   Knowledge based systems analysis and design
VERYARD   Information modelling: practical guidance
VERYARD   Information coordination: the management of information models, systems and organizations
WELLMAN   Software costing

Analysing Systems

Determining Requirements for
Object-Oriented Development

Roy MacLean, Susan Stepney, Simon Smith,
Nick Tordoff, David Gradwell, Tim Hoverd
and Simon Katz

Prentice Hall
New York   London   Toronto   Sydney   Tokyo   Singapore

First published 1994 by
Prentice Hall International (UK) Limited
Campus 400, Maylands Avenue
Hemel Hempstead
Hertfordshire, HP2 7EZ
A division of
Simon & Schuster International Group

© Logica Cambridge Ltd and Data Dictionary Systems Ltd

All rights reserved. No part of this publication may be reproduced,
stored in a retrieval system, or transmitted, in any form, or by any
means, electronic, mechanical, photocopying, recording or otherwise,
without prior permission, in writing, from the publisher.
For permission within the United States of America
contact Prentice Hall Inc., Englewood Cliffs, NJ 07632

Printed and bound in Great Britain at
the University Press, Cambridge

Library of Congress Cataloging-in-Publication Data

Available from the publisher

British Library Cataloguing in Publication Data

A catalogue record for this book is available from
the British Library

ISBN 0-13-301433-9

2 3 4 5   98 97 96 95 94

Contents

Preface   xi

I   Introduction to ORCA   1

1   Why ORCA?   3
    1.1   The ORCA method   3
    1.2   Analysis and design   3
    1.3   An analysis method   5
    1.4   Other OO methods   7

2   ORCA by example   9
    2.1   Introduction   9
    2.2   Preliminary analysis   9
    2.3   Finding out what's wrong   11
    2.4   Prescribing a remedy   14
    2.5   Specifying a solution   15
    2.6   Change and development   24
    2.7   Conclusion   24

3   ORCA's key ideas   25
    3.1   An analysis process needs to be tailorable   25
    3.2   A system has purposes as well as behaviour   25
    3.3   A system is a network of co-operating roles   26
    3.4   A system may exhibit pathologies   27
    3.5   Models can describe the Old World, or a New World   28
    3.6   A system is embedded in an environment   29
    3.7   IT systems are ways of implementing behaviour   30
    3.8   Models can express different levels of abstraction   30

Contents

    3.9   Overview of ORCA   32

II   The Basic Process   35

4 Introduction to the case study   37
    4.1   Introduction to Part II   37
    4.2   Overview of the ORCA Basic Process   38
    4.3   Brief description of the case study   40

5 Preliminary Analysis   45
    5.1   Introduction   45
    5.2   Initial understanding of the client's world   45
    5.3   Establish the terms of reference   46
    5.4   Identify sources of information   50
    5.5   Define the contractual task   50

6 Process Design   52

7 Old World Purpose   54
    7.1   Introduction   54
    7.2   NIMWeC as a manufacturing organisation   55
    7.3   Investigating NIMWeC   59
    7.4   Further investigation   66

8 Old World Behaviour   71
    8.1   Introduction   71
    8.2   The weaving of name tapes   72
    8.3   Orders, batches and the batching process   82
    8.4   From orders to name tapes   85

9 Pathology and Prescriptions   89
    9.1   Introduction   89
    9.2   Adding in the development objectives   90
    9.3   Structuring pathologies, positing prescriptions   91
    9.4   Evaluating the prescriptions   97

10 Specifying the New World   103
    10.1   Introduction   103
    10.2   What the New World looks like   104
    10.3   Onward into development?   112

III Using the Modelling Languages   113

11 Purpose and Behaviour   115
   11.1 Why use models?   115
   11.2 Purposive entities and Behavioural entities   116

12 Behavioural modelling   121
   12.1 Histories and Frameworks   121
   12.2 Object characterisation   124
   12.3 Dynamic behaviour   129
   12.4 Levels of abstraction   140
   12.5 The Class and Framework constructs   146
   12.6 Structural frameworks and temporal frameworks   147
   12.7 Behaviours and Services   149

13 Do's and don'ts   150
   13.1 Introduction   150
   13.2 Process   150
   13.3 Modelling   151
   13.4 Diagrams   151
   13.5 Abstraction   152
   13.6 Classification and specialisation   152
   13.7 Naming   153

IV Tailoring the Process   155

14 Introduction to tailoring   157
   14.1 Why do we need to tailor the process?   157
   14.2 Overview of tailorings   157

15 Organising the organisation—National Parks   159
   15.1 Analysis parameters   159
   15.2 Preliminary analysis   160
   15.3 Modelling the Old World   162
   15.4 Determining the system pathology   164
   15.5 Drawing conclusions—the Analysis Report   167

16 Shaking up the business—Just in Time   169
   16.1 Analysis parameters   169
   16.2 Preliminary analysis   170
   16.3 Describing the Old World   171
   16.4 Determining the Pathology   173
   16.5 Specifying a New World   174
   16.6 Prescribing Change   176

16.7	Behavioural requirements	177
16.8	Development and transition	180

17 A new Purpose in life—the Paperless Map — 182
17.1	Analysis parameters	182
17.2	Preliminary analysis	183
17.3	New World behaviour	184
17.4	Development strategy	185
17.5	Enhancing our kitbag	186

18 No Old World—spreadsheets and telephony — 188
18.1	Analysis parameters	188
18.2	Why omit the Old World?	188
18.3	Modelling or design?	190
18.4	No Old World processes	191
18.5	Summary	192

19 Nothing new under the sun—a lending library — 194
19.1	The analysis strategy	194
19.2	A lending library	195
19.3	Other uses, other models	202

20 No-one to talk to—Ahab applied to NIMWeC — 208
20.1	The need for tailoring	208
20.2	Balderdash	208
20.3	Ahab	209
20.4	Applying Ahab to NIMWeC	210
20.5	Outputs from Ahab	214
20.6	Adding Ahab to the analysis process	216

V Wider Issues — 219

21 Putting together an analysis project — 221
21.1	Planning, monitoring and estimation	221
21.2	Validation	222
21.3	HCI requirements	223
21.4	'Non-functional' requirements	223
21.5	Configuration management	226
21.6	Tool support	226

22 Life after ORCA—onward into development — 230
22.1	Design and implementation considerations	230
22.2	Design methods	232
22.3	The ORCA method	234

VI Appendices   235

A Defining modelling languages   237
A.1 Introduction   237
A.2 Syntax   237
A.3 Semantics   240

B Grampus—the Purposive modelling language   242
B.1 Introduction   242
B.2 Roles and services   242
B.3 Cluster and co-operation   244
B.4 Formation, delegation and promotion   247

C Beluga—the Behavioural modelling language   250
C.1 Beluga models   250
C.2 Classes   250
C.3 Frameworks   255
C.4 Framework statics   256
C.5 Framework class relationships   259
C.6 Framework dynamics   260
C.7 Diagrammatic concrete syntax   267
C.8 Generalisation and abstraction   280

D Glossary   283

E Bibliography   289

F Index   291

Preface

About this book

This book describes and explains ORCA (Object Oriented Requirements Capture and Analysis). It is aimed at all those interested in the description and analysis of complex systems such as businesses, social organisations, and so on. In particular, it is aimed at those responsible for producing strategies for the use of Information Technology (IT) and requirements for software development.

Although we refer to ORCA as a 'method', this book is not a manual to be followed slavishly during the analysis process. The usefulness of such a manual would in any case be questionable. Rather, this book presents and explains a set of ideas about analysis, together with ways of using these ideas. We hope that these ideas will be useful to the practising analyst. Consequently, the ideas are presented primarily through worked examples, rather than in a generalised, 'text book' manner. We believe that this illustrates better how the ideas can be used in practice, and avoids the impression of a rigid, prescriptive method.

Producing and analysing models is a primary activity in ORCA and two complementary modelling languages are provided for this purpose. These languages are illustrated by example models throughout the book. In addition, formal syntax definitions, together with explanations of each language construct, are provided in appendices.

The other aspect of ORCA is its view of the analysis process. This is presented as a basic process together with a number of process variants that are appropriate in different analysis situations. ORCA includes a stage where the analyst should decide on the kind of process required for the given analysis situation. The different processes illustrated in the book (using various case studies) can be used to help in this choice of approach.

One of the example analyses (the weaving factory, in Part II) is considerably more substantial than the others. This is intended to illustrate the analysis of a system that is complex and unfamiliar—the very situation for which we want an analysis method. Using only 'toy' examples would not convey the difficulties encountered (and, we hope, overcome) in real analyses.

Overview of the book

In Part I we introduce ORCA; this part can be read as a stand-alone overview of the method. Chapter 1 sets the scene, and provides the motivation for developing an analysis method, and for ORCA in particular. Chapter 2 illustrates ORCA using a simple example based on a petrol filling station. The example is not intended to be realistic either in scale or level of detail. However, most of the features of ORCA are covered. Chapter 3 summarises the key ideas of ORCA, including some that are not illustrated fully in Chapter 2.

In Part II we illustrate ORCA's Basic Process in a weaving factory case study. The various chapters introduce the main ORCA concepts in the context of the example; they should be read in sequence. Chapter 4 introduces the weaving factory case study, and gives an overview of the Basic Process. Chapter 5 concerns Preliminary Analysis—gaining an initial understanding of the client's world, establishing the terms of reference of the analysis, and identifying sources of information. Chapter 6 concerns the activity of deciding on an appropriate analysis process; in this case, the Basic Process is assumed to be adequate. Chapters 7 and 8 develop and analyse models of the Old World—the existing situation with the weaving factory. Chapter 9 discusses pathologies (what is fundamentally wrong with the current world) and prescriptions to remedy these. Chapter 10 specifies a New World for the weaving factory—this includes both organisational changes and IT developments.

In Part III we show, in some more detail, how the modelling languages can be used. Chapter 11 discusses the difference between purposive and behavioural models; Chapter 12 discusses behavioural modelling in depth; Chapter 13 offers some general heuristics on using ORCA.

In Part IV we show how ORCA's Basic Process may be tailored to suit different analysis situations, using a variety of example case studies. Many of these are inspired by real-world examples, but have been modified for explanatory purposes. These examples can be used to provide a starting point for a specific tailoring. Chapter 14 introduces the idea of tailoring the Basic Process. Chapter 15 considers the analysis of a loose organisation of semi-independent units—in this case, a federation of National Parks trusts. Chapter 16 looks at an example of radical restructuring of a manufacturing business—a change from traditional stock control to 'just-in-time' delivery. Chapter 17 examines the case where the overall mission of an organisation changes and the existing world has to be re-engineered in the light of the new purposes. Chapter 18 considers analysis situations where there does not seem to be an Old World, because a radically new product is being developed, or the world is developing into new areas. Chapter 19 considers an analysis process that involves reuse, by adapting pre-existing generic models, rather than constructing bespoke models from scratch. Chapter 20 provides a technique for analysis of textual documents, for use when these are the only source of information for an analyst. Chapters 15–20 are independent of each other.

In Part V we discuss wider issues concerning ORCA analysis. Chapter 21 considers issues such as planning and monitoring, validation, requirements for human–computer interaction, 'non-functional' requirements, configuration management and tool support. Chapter 22 discusses how to proceed into software design and development after an ORCA analysis has concluded.

In Part VI we give a more formal description of ORCA's modelling languages. This part is for reference, and is not intended to be read sequentially. Appendix A describes how we go about defining the modelling languages. Appendix B defines Grampus, the purposive modelling language, and Appendix C defines Beluga, the behavioural modelling language. These appendices are intended for those interested in delving deep into the corners of the modelling language, for example in order to build tools. Appendix D is a glossary of ORCA terminology and other supporting terms.

The ORCA project

The ORCA method was developed during a three-year project that took place between mid-1990 and mid-1993. This project was part-funded by the United Kingdom's Department of Trade and Industry (DTI) and Science and Engineering Research Council (SERC) under the IED programme (project number IED4/1/2134); this support is gratefully acknowledged.

The project collaborators are Logica UK Ltd, Data Dictionary Systems Ltd, and the Department of Computer Science, University of York.

We would like to thank Roy Flowerdew of Wovina Woven Labels for permission to use the case study contained in Part II of this book. Unlike many other texts on systems analysis, this case study represents a real problem for which a real implementation has been carried out. Some specific features were enhanced for use in this book, but the spirit of the example is very much in line with the Wovina system and its implementation.

We would also like to thank Simon Bennett, Alan Brown, Peter Hitchcock, Jennifer Stapleton and Dave Whitley for their ideas, comments and contributions during the course of the project.

Part I

Introduction to ORCA

Chapter 1
Why ORCA?

1.1 The ORCA method

ORCA aims to be an effective, flexible and usable method for requirements capture and analysis, based on object oriented concepts. ORCA defines and supports an *analysis process*, provides appropriate *modelling languages*, and provides guidance concerning their use.

ORCA's preferred mode of use is for performing analysis that leads into object oriented (OO) design and implementation using existing methods. ORCA can also be used where Information Technology (IT) development will not—for whatever reasons—use OO technology. It can even be used for analysis where no IT development is foreseen. In fact, an analysis should never start with the assumption that IT development will take place (with all due respect to those clients who 'want to computerise X').

Successful analysis is difficult, and no method can take the practitioner mechanically through the activities involved—there is no substitute for experience and insight. However, ORCA attempts to provide both conceptual and practical equipment for carrying out the analysis task, building on existing ideas and methods.

1.2 Analysis and design

1.2.1 Why are we interested in analysis?

The purpose of analysis is to provide a context, a rationale, and a specification for change to part of the world. Informally, we analyse a delimited and coherent portion of the world, such as a business, a manufacturing or control process, or a social organisation. We often refer to such a portion of the world as a 'system'.

Change may or may not involve the development and installation of IT components: computer hardware and software, communications links, and so on. In this book, we assume a general intention to carry out IT development, but this does not affect our approach to analysis. It is a fundamental principle that, where possible, analysis should question the need for IT development.

Where IT developments are being carried out, both the client and the developer have an interest in knowing the following:

- Why is the development taking place?
- What problems are the IT components going to solve?
- What, in general terms, are they going to do? Why do we think this is going to solve the problems?
- What changes to the client's world are necessary, other than the development and installation of IT components?

These are the questions the the analysis activity should address. In some cases, the answers may seem to be obvious, or they may be given as unquestionable dictates. The dangers of either situation should be readily apparent. We should seek to pursue an analysis process.

1.2.2 The analysis process

The analysis process needs to take us through the following tasks:

- determining the scope of the analysis—what is relevant, and what isn't
- identifying problems and diagnosing the underlying causes
- proposing remedial changes, and justifying why we believe they are going to have the desired effect
- specifying IT components that implement the changes, and saying why these are necessary to support the remedial changes

The specification of each IT component has two aspects:

- the changed world's view of the IT component—who or what interacts with the IT component? when? how?
- the IT component's view of the world—what does it need to 'know' about the outside world? what events does it need to respond to?

A fundamental tenet of ORCA is that such specifications should be produced as part of an analysis process that also provides a context and a rationale for development. This process cannot assume any 'initial inputs', such as an informal statement of requirements—it must proceed from the earliest familiarisation with a client's world. The ORCA process therefore seeks to link Business Analysis with System Development.

1.2.3 Analysis versus design

Once we have an initial specification of an IT component, we are into the process of design and implementation. There is a fundamental difference between analysis and design that is worth a short digression.

Analysis is concerned with what is *essential*—what is essential to the client's world and its environment, what it is necessary to change, what we need to say about how the changed world behaves. In contrast, design is concerned with what is *adequate*—what mechanisms for information handling and interaction with the outside world are adequate for the task being supported, what level of programming will ensure that a component works in the intended manner.

If we were talking about office buildings, say, the distinction would be obvious: on the one hand the descriptions of the essential ambient conditions, and on the other hand the air conditioning system that is adequate to ensure these conditions, are clearly different things. In contrast, software components are expressed, throughout their development, using 'artificial' specification and programming languages. Because the same languages may also be used as tools for thought within the analysis process, the distinction between the products of analysis and the products of design is not so clear.

This has benefits and dangers. One benefit is that similar concepts serve for both analysis and design—no fundamental change of mental apparatus is required. Another benefit is that the models built up during the analysis process can potentially provide material inputs to the design process. Existing methods for object oriented development place considerable emphasis on this—the 'information model' that results from analysis defines the principal classes within the software, often with little or no adaptation. The danger is that the major design decision involved in this step passes unnoticed and unquestioned. We therefore need to be very careful about the interpretation of models as analysis models or design models, especially when they use similar languages.

1.3 An analysis method

1.3.1 Requirements of an analysis method

An analysis method should support the *basic process* set out in Section 1.2.2:
- determining the scope of the analysis
- identifying problems and diagnosing the underlying causes
- proposing remedial changes
- specifying IT components that implement the changes

The principal technique supporting this process is the construction and analysis of models. An analysis method should therefore provide one or more *modelling lan-*

guages, together with guidelines for their use. The nature of the analysis activity gives rise to several requirements for these modelling languages. These requirements are outlined below; Chapters 11 and 12 provide details of ORCA's response to them.

When we start to analyse the world, we do not, in general, know what level of abstraction we need to deal with—if we are told, we should question the assertion. For example, is the problem with the Wonder Widget Company in its relationship with its customers, or with the way its divisions are organised, or with the widget manufacture process, or with the efficiency of the widget-pressing machines, or...? Finding the right level of abstraction for expressing problems and solutions is an essential aspect of analysis. Consequently, our modelling languages need to offer facilities for expressing and relating *descriptions at different levels of abstraction*.

Our modelling languages should allow us to distinguish between what people believe ought to be happening in the world, and what actually does happen (at present or in some future changed world). These can be seen as *descriptions of purpose and behaviour* respectively. We will have more to say about this distinction in subsequent chapters.

In analysis we cannot assume that we have a single, reliable source of information. In general, we will want to capture *multiple local viewpoints*. This is particularly relevant to descriptions of purpose, since different people can hold different, and perhaps inconsistent, views of what a given component is for, even if they agree on what it actually does. On the other hand, incompatible views of behaviour are resolvable, in principle if not in practice, by further information gathering or observation.

Another 'fact of life' for analysis is that behaviour happens over time and is inherently concurrent—different strands of behaviour can be going on at the same time. It is often important in analysis to describe when certain events happen, relative to other events, and how concurrent strands of behaviour interact. We therefore need a modelling language that allows us to express and analyse *dynamic behaviour*.

As the discussion in Section 1.2.3 indicates, analysis is concerned with what is essential regarding change to the world. The analyst should seek to give the IT developer the loosest specification that will result in an acceptable product. In other words, an analyst should not make 'premature design decisions' that would restrict the developer's options. We therefore need a modelling language that allows us *controlled 'don't care' lack of constraints*, specifically in describing behaviour. The opposite is the case in the design process—we need eventually to be as precise as possible, explicitly making decisions between implementation alternatives.

1.3.2 Why object orientation?

Many of the approaches to IT systems development in use today have their roots in the times when computing power was expensive. Then, it made sense to have a 'system-centred' architecture, where a system's behaviour is specific for a particular business situation, and the system drives the user's responses.

Now, the thrust of systems development is towards 'user-centred' systems. This is most obvious in the area of graphical user interfaces (GUIs), which put more freedom and responsibility into the hands of the users—allowing them to control the system and exercise their skills more effectively. User-centred systems are typically structured as a number of software components that respond to specific user events, the sequence and structure of which is necessarily unknown to those components.

Object orientation (OO) is a particularly appropriate technology for the development of user-centred systems, since a collection of objects responding to user events, through the mechanism of the operations provided by those objects, is exactly the structure required. A 'client–server' approach to systems development has all of the components of a system, not just the user interface, structured as components that respond to external requests. The client of a particular server is the 'user' of that component. Hence, OO is a key technology in the development of modern, responsive computer systems.

Developing these more flexible systems carries with it a new problem. With traditional computer systems, the computer processes are driving all of the system's activity, so everything about the system is in the domain of the development method, and thus it is possible to have a prescriptive development method. User-centred systems do not have this property because they are directly responsive to a wider range of input stimuli—the user of a particular component might do anything next. It is no longer possible to define significant completeness and consistency rules for a development method. This book presents a flexible approach to requirements definition for user-centred systems.

ORCA's modelling languages, defined and used in the rest of this book, provide the analyst with a way of looking at and talking about the world in terms of system components that respond to user-originated events. They also allow for the definition of a system's requirements in terms of the functionality that is delegated, as a collection of available services, to various system components.

Transition from analysis into design should be easier if they share a common conceptual basis. We therefore need an analysis method based on object oriented concepts, in order to complement object oriented development approaches. Even if the software technology used for IT development is not object oriented, the usefulness of OO for modelling makes it worth while as the basis of an analysis method.

1.4 Other OO methods

There are various methods for 'object oriented analysis and design'. Broadly speaking, these other methods do not fully support the analysis process as described in

Section 1.2.2. In particular, the idea that IT development is part of wider change is not explicit in most of these methods. An ORCA analysis encompasses the environment and context, and may not identify which components, if any, are to be the subject of IT development until relatively late in the process.

Some methods presuppose the existence of a 'requirements specification' document or 'problem statement'—the precise term varies—as the starting point for development. Such a document describes the required IT system informally. 'Analysis' is then largely a formalisation of the requirements specification, not an analysis of the real world. For example, noun phrases within the specification can be used to suggest candidate classes in an OO model. This model is a primary input to the design process. However, much work may be needed to produce a good 'requirements specification', which is, by definition, not covered by these methods. In any case, a developer is unlikely to receive from a client a requirements specification that is detailed and comprehensive enough to act as the sole source of information, so the developer will still need to carry out 'real', rather than documentary, analysis. While a textual statement of requirements might be needed for contractual purposes, it seems sensible to derive this from the analysis models, not the other way round. A significant part of the ORCA process is dedicated to the analysis necessary for eliciting requirements.

A further problem with existing OO methods' support for analysis is that traditional object orientation, on which these methods are based, does not fully satisfy the requirements outlined in Section 1.3.1. ORCA goes beyond these other methods by providing the additional support needed to satisfy these requirements.

Chapter 2
ORCA by example

2.1 Introduction

In this chapter we illustrate ORCA using a simple example based on a petrol filling station. The example is not intended to be realistic either in scale or level of detail. Consequently, the analysis process and the models used within it are presented only in outline. Chapter 3 summarises the key ideas of ORCA, including some that are not illustrated fully in this chapter. In Part II we describe ORCA's basic process in greater detail, using a much larger example.

2.2 Preliminary analysis

We have been called in by the owner of a petrol station because of poor sales. Our client believes that customers are going elsewhere because of the unacceptable frequency with which pumps have to be taken out of service while waiting for deliveries of petrol from the supplier.

Our first activity is to conduct a preliminary analysis, in which we get our bearings on the problematic situation, and determine the scope of the analysis. No formal modelling is involved at this stage.

Currently, orders to the supplier are made on a regular basis (weekly, at the moment), but the variability of demand means that stocks of petrol often run out before the next delivery. When this happens, the ordering cycle can be brought forward, but there are inevitable delays before the petrol is received. Since the supplier makes a charge for each delivery, it is not acceptable simply to increase the frequency of orders.

We summarise the situation as the 'rich picture' shown in Figure 2.1.

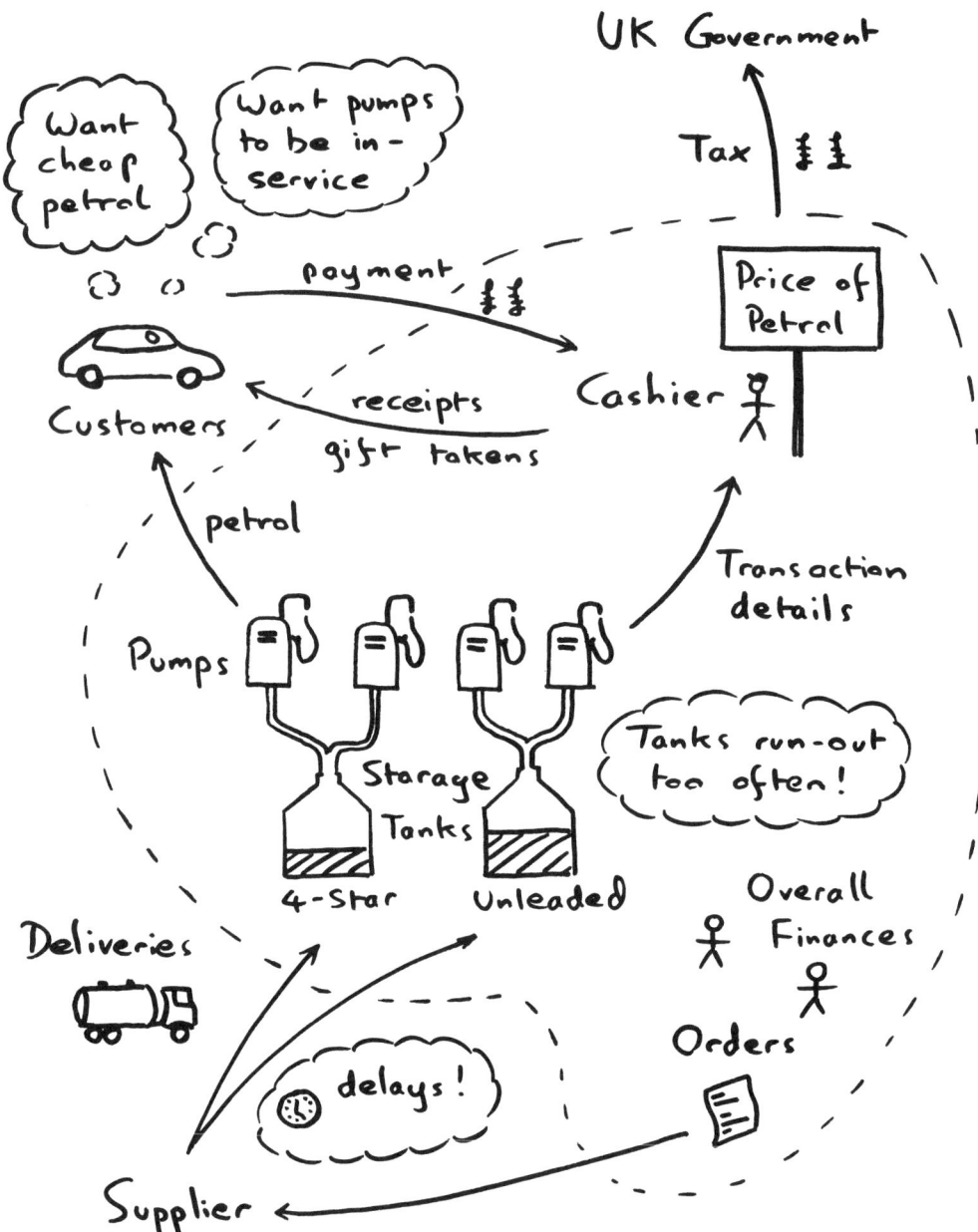

Figure 2.1 Petrol Station rich picture

2.3 Finding out what's wrong

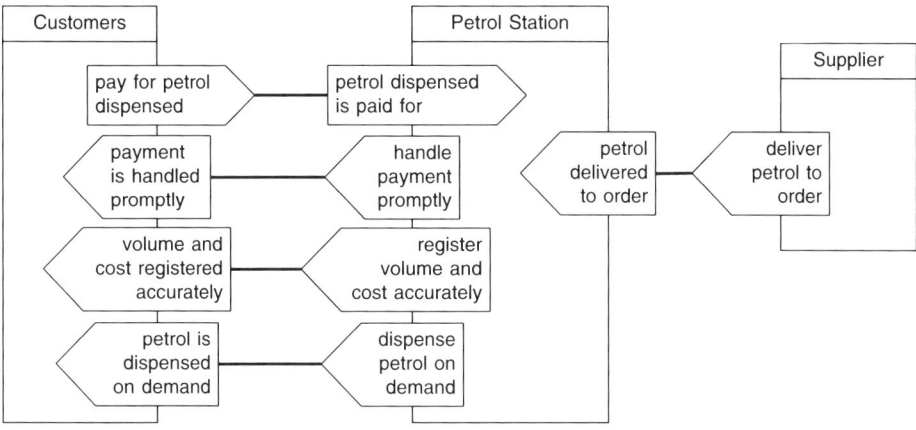

Figure 2.2 Service provision relationships

2.3 Finding out what's wrong

In order to find out what is wrong, we first need to establish what is essential to the system—what it is 'about'. In an ORCA analysis, a system is viewed as a network of *service provision* relationships: both the arrangements that exist with regard to service provision, and the actualities of service provision.

In the case of the Petrol Station, the system is in a typical customer–retailer–supplier situation. The essential services seem to be as follows:

- The Petrol Station provides petrol to Customers on demand.
- Customers pay Petrol Station for petrol dispensed/received.
- Petrol Station registers the volume and cost of petrol dispensed accurately, and communicates this to Customers.
- Petrol Station handles Customers payment promptly.
- Supplier delivers petrol to Petrol Station as per order.

The first four of these services relate to transactions with Customers, and the fifth to transactions with the Supplier.

We can fairly say that all these services are essential to the petrol station system. If any of them are not agreed to by the parties concerned, or if the services do not happen behaviourally, then the system as a whole will not 'work'.

Figure 2.2 shows the service provision relationships between Customers, Petrol Station and Supplier. The three *roles* in the system are shown as rectangular boxes, and the service provision relationships as links between these. Each service provision relationship has a 'guarantee' end and a 'rely' end. A *guarantee* is shown as a smaller pointed box pointing out of a role box; a *reliance* is shown as a smaller pointed box pointing into a role box. The text within a pointed box is a

12   Chapter 2   ORCA by example

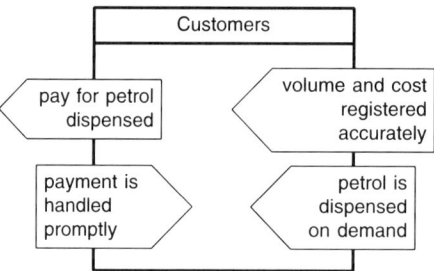

Figure 2.3 Customers' view of services

description of the service involved, from either the active (guarantee) or passive (rely) viewpoint.

For example, Customers guarantee to pay the Petrol Station for petrol dispensed/received; the Petrol Station relies on this being the case. At least, this is the essential situation—an assumption on which the whole system is founded. In practice, this arrangement might not work, due to large numbers of absconding customers, and this would clearly be a problem. For the purposes of our example analysis, we assume that this is not the case.

Each of the three roles has its own view of the world, in terms of the services that it is supposed to provide to other roles, and the services that other roles are supposed to provide to it. For example, Customers expect petrol to be provided on demand, expect the pumps to be accurate, expect prompt handling of their payments, and agree to pay for petrol received (Figure 2.3). A minimum requirement for the system to work properly is that all the local views of service provision are consistent. For every role that relies on a service, some other role (or group of roles) should guarantee to provide it. Such roles are then in a *co-operation* with respect to that service.

In our model, the co-operations between the three roles seem to be valid, but in practice we would like more precise definitions of the *service qualifiers*—'promptly', 'accurately', 'on demand', and so on. For example, the petrol station cashier might interpret 'promptly' to mean 'when I've finished my cup of tea', whereas the customers might prefer it to mean 'within a few seconds'.

An important point to note is that a co-operation is not a 'flow' of physical items or information. Co-operations are requirements placed on behaviour; behaviour that is deemed to realise a service may sometimes fall into a 'producer–item–consumer' pattern, but this is not necessarily the case. For example, Petrol Station provides the service 'handle payment promptly' to Customers, even though money (in some form) 'flows' from Customers to Petrol Station. As we noted above, there is also a qualitative aspect to service provision: timeliness, accuracy, availability, and so on.

In the preliminary analysis we recorded the client's informal problem description. This suggested that the problem of falling sales was due to the frequency with which pumps are out of service while waiting for deliveries of petrol from the Supplier. Using the model we have built, we can check this hypothesis.

Sales are falling because customers are switching to other petrol stations in the area. So, presumably, some of the services that they rely on from the Petrol Station are not being provided adequately. The client assures us that neither payment handling nor the accuracy of the pumps is at fault, so the 'provide petrol on demand' service must be deficient.

In order to find out why this is so, we need to know more about the Petrol Station. We do this by finding out how the reliances and guarantees of the Petrol Station as a whole correspond to the reliances and guarantees of 'subroles' within it. Our investigations lead us to suggest the following subroles:

- Petrol Dispensing—providing petrol on demand and registering volume and cost accurately
- Payment—requiring and handling customer payment
- Ordering—requiring delivery of petrol to order

In addition to the reliances and guarantees that correspond to those of the Petrol Station role, the following co-operations link these three roles:

- Petrol Dispensing guarantees provision of transaction details (volume and cost) to Payment, so that appropriate payment can be requested from Customers.
- Ordering guarantees availability of petrol to Petrol Dispensing.

The relationship between Petrol Station and its subroles is shown in Figure 2.4. The subroles of Petrol Station identified above, and the co-operations between them, are a rationalisation of the petrol station system, in terms of service provision. They are *not* a model of its organisational structure. If the petrol station is 'well organised', there should be a close correspondence between the roles and co-operations in our model, and the actual organisational structure. However, it is quite likely that the petrol station has a different organisational structure, or perhaps no explicit organisation at all (a handful of employees doing various jobs as necessary).

We can now return to the problem analysis. Given that the 'provide petrol on demand' service is deficient, either Petrol Dispensing is at fault (for example, the pumps keep breaking down), or the reliance on Ordering is not being satisfied. Again, the client assures us that the pumps are reliable, so the problem lies with the availability of petrol. Either there is something wrong with the way that ordering is done, or the supplier does not deliver to order. Since the performance of the supplier is not at fault, we conclude that the ordering behaviour is the source of the problem.

Notice that the co-operation between Petrol Dispensing and Ordering is valid: there is agreement about what 'ensure availability of petrol' means. In this case, it means that the pumps should never be out of service through lack of petrol; deliveries should be made before stocks (in the storage tanks) are exhausted. The

14   Chapter 2   ORCA by example

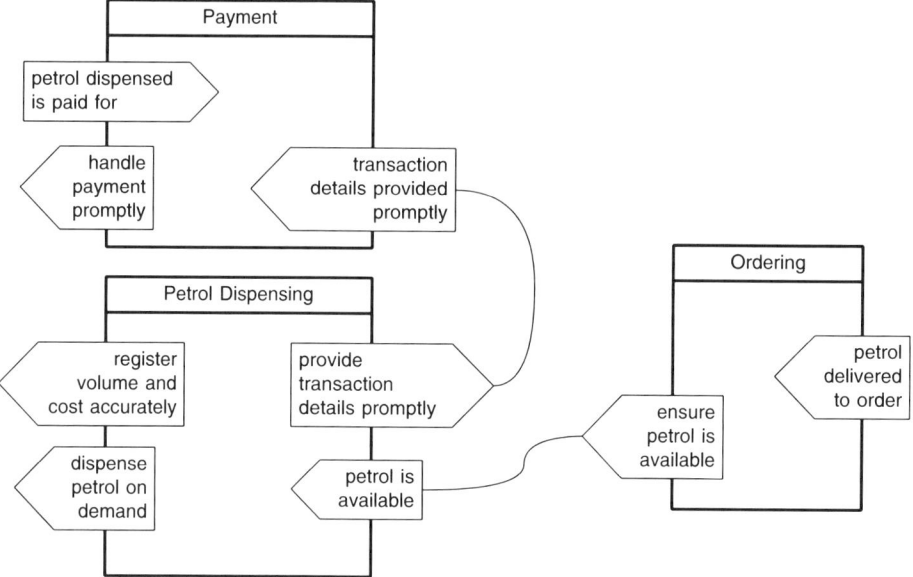

Figure 2.4 Petrol Station and its subroles

problem lies with the way in which the service is currently *realised*. It just doesn't work well enough.

Having localised the problem, we now need to examine the deficient behaviour—the current procedures for monitoring stock levels and generating orders. In this case, the current behaviour is simple enough not to require explicit modelling. The essence of the problem is that a regular ordering cycle is too crude a strategy to cope with the existing variability in demand.

2.4   Prescribing a remedy

At this point in the analysis process a combination of insight and experience is required. No method can generate solutions mechanically.

Currently, the ordering behaviour takes no account of the rate at which petrol is being dispensed, or the cumulative quantity dispensed. Given this information, it is feasible to predict when a storage tank will be exhausted. If the typical time between placing an order and receiving the delivery is known, orders could be generated pre-emptively. The aim might be for deliveries to arrive when the storage tank concerned is 5 per cent full (thus allowing a suitable margin for error).

As it happens, Petrol Dispensing already guarantees to Payment that information about transactions is provided. This information includes the grade of petrol and the volume dispensed. In addition, the payment handling behaviour records

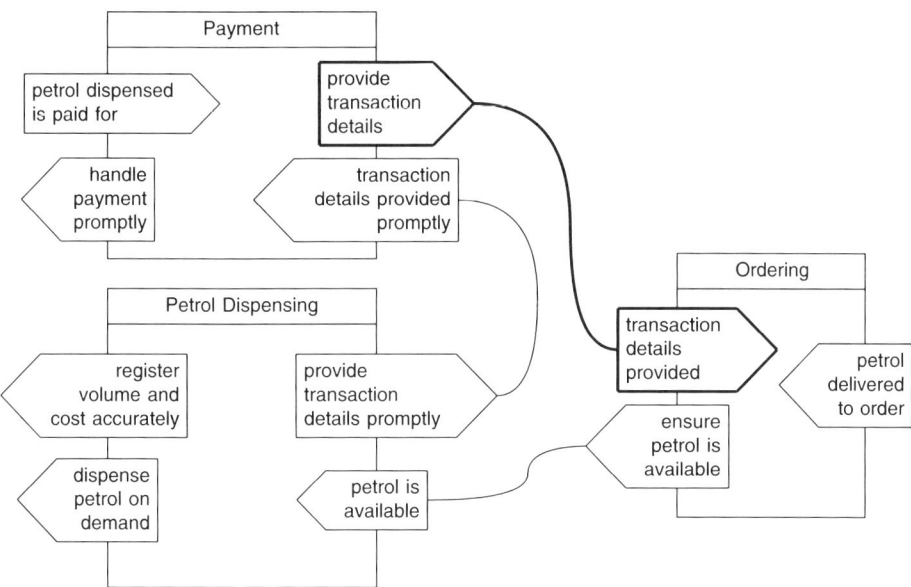

Figure 2.5 The prescribed remedy. The additional co-operation is shown in bold outline

the date and time of the transaction. If this information is then made available to the ordering behaviour, there is a basis for predicting future consumption. We simply need an additional co-operation between Payment and Ordering (Figure 2.5). Notice that Ordering does not require the transaction information promptly, unlike Payment. To be more precise, Ordering needs the information 'promptly' relative to the timescale of ordering petrol (days), rather than relative to the time that customers are happy to wait (seconds).

What Ordering needs to do is to monitor consumption and generate orders at appropriate times (rather than on a regular basis, regardless of consumption). It can do this on the basis of the transaction information (and perhaps other information as well). The quantity and complexity of information involved in this task makes IT support desirable. We therefore propose to our client that

- a new, more sophisticated ordering strategy be adopted
- the ordering activity be automated

2.5 Specifying a solution

The next stage of the analysis is to model those aspects of behaviour that are relevant to our new ordering strategy. Aspects that are not relevant (for example, to do with customer payment) are not modelled. The model identifies the kinds

16 Chapter 2 ORCA by example

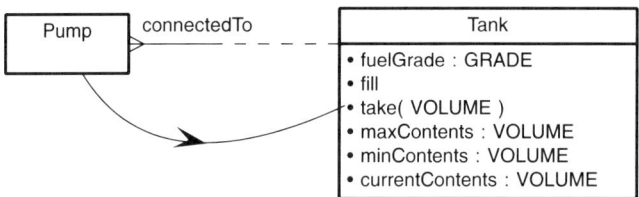

Figure 2.6 Pumps and Tanks

of object involved in the behaviour and their inter-relationships (*statics*), and the patterns of events within the behaviour (*dynamics*). This model forms the basis for a statement of requirements with regard to IT development. It is a *behavioural model*, using ORCA's Beluga language. In contrast, the earlier models, which deal with views of service provision, are *purposive models*, using ORCA's Grampus language.

The distinction between these two kinds of model is fundamental to ORCA. A purposive model describes the way that roles within a system are linked by arrangements to co-operate with respect to particular services. A behavioural model describes the objects and events that exist (or need to exist) in order to realise the services.

At this stage of analysis, we do not distinguish between 'real world' objects and software objects within some proposed IT system. It is for the statement of requirements, ultimately produced by the analysis, to make clear which elements of our models need to be handled by IT components. It is also worth pointing out that we could have used behavioural modelling in the 'finding out what's wrong' stage, in order to investigate the original deficient behaviour, but this was not necessary in our simple example.

2.5.1 Static structure

First, we need to characterise the *objects* involved in the behaviour, in terms of their *classes* (types of object) and *features* (attributes of, and operations on, objects of a particular type). In looking for behavioural entities, we need to go back and examine the service descriptions (the reliances and guarantees) in our purposive models.

Basic physical objects, such as pumps and storage tanks, are a good place to start our modelling (Figure 2.6). The diagram says that each *Pump* must be connected to exactly one *Tank*, and each *Tank* may (dashed line) have any number of *Pump*s connected to it (indicated by the *crows-foot*). The features *maxContents* and *fuelGrade* are fixed attributes of a *Tank*; *currentContents* is a variable attribute. Note that the type of an attribute is given after the colon. There

2.5 Specifying a solution   17

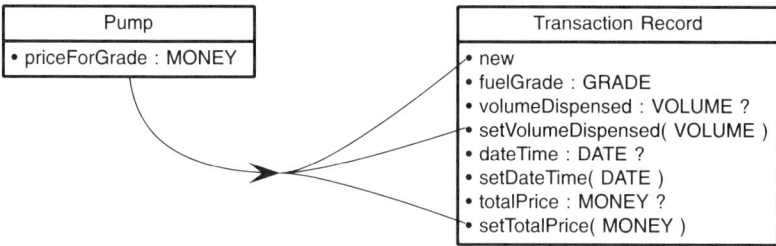

Figure 2.7 Transaction Record

is a constraint (not shown in the diagram) that *currentContents* must always be less than or equal to *maxContents*:

  invariant : $currentContents \leq maxContents$

The features *fill* and *take* are operations that change *currentContents*. The effect of a *fill* operation is to make *currentContents* equal to *maxContents*:

  fill   //guarantee : $currentContents = maxContents$

The *take* operation must be given a VOLUME as a parameter, and decreases *currentContents* by the specified amount:

  take(v)   //guarantee : $currentContents' = currentContents - v$

(*currentContents'* refers to the value of the attribute after the operation has been performed.) This operation can be invoked by any connected *Pump* (curved arrow).

The information concerning a transaction is modelled as a *TransactionRecord* (Figure 2.7). A *TransactionRecord* is created by a *Pump* every time it is used (shown in the diagram by the curved arrow to the *new* operation). The *fuelGrade* is that of the *Tank* to which the *Pump* is connected. The *volumeDispensed* and the *totalPrice* of the dispensed petrol are recorded by the *Pump* (shown by the curved arrows to the various *set* operations). The *dateTime* of a *TransactionRecord* is set within *Payment* before *Ordering* receives the *TransactionRecord*. Notice the convention whereby the attributes such as *volumeDispensed* are not directly modifiable—they can be changed only by the appropriate *set* operation. The attributes of a *TransactionRecord* may or may not be set to some value, depending on where it is in its *life cycle*; this is indicated by the '?' suffix, indicating optionality.

The behaviour also deals with *Orders*. These are created by *Ordering*, and communicated to *Supplier*. Some time after an *Order* has been placed, *Supplier* fulfils the order by making a delivery; the *complete* operation on an order records delivery details using *deliveryDate* and *volumeDelivered* (Figure 2.8). The diagram says that each *Order* must be made to a single *Supplier* (we know that there is

18 *Chapter 2 ORCA by example*

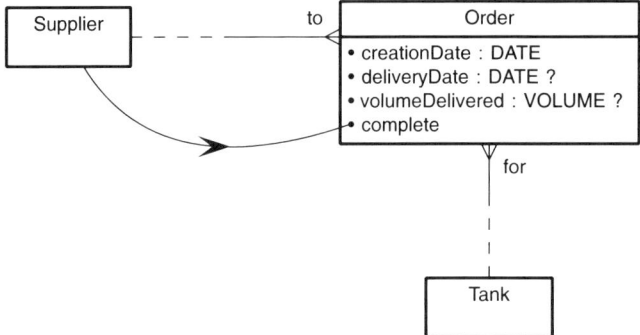

Figure 2.8 Orders

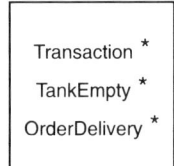

Figure 2.9 Dynamics

only one), and must be for a single *Tank*; *Supplier*s and *Tank*s may be associated with many *Order*s. The *complete* operation is instigated by *Supplier* when they make a delivery. The *deliveryDate* and *volumeDelivered* attributes are set by the *complete* operation, some time after creation of an *Order*. Thus, these attributes are initially not set, and this is indicated by the '?' suffix.

2.5.2 Dynamics

The dynamics of the ordering-related behaviour involves three kinds of *episode* (Figure 2.9):

- *Transaction*: transactions in which pumps take petrol from tanks, for dispensing to customers
- *TankEmpty*: tanks running low on petrol, causing the connected pumps to be taken out of service
- *OrderDelivery*: orders for more petrol, and deliveries by the supplier

There are many occurrences of each kind of episode, indicated by the '*'s after the episode name. The occurrences can, in general, happen concurrently. For example, *OrderDelivery* cycles will be going on while petrol dispensing *Transaction*s are happening. However, as we see below, the different kinds of episode are not totally

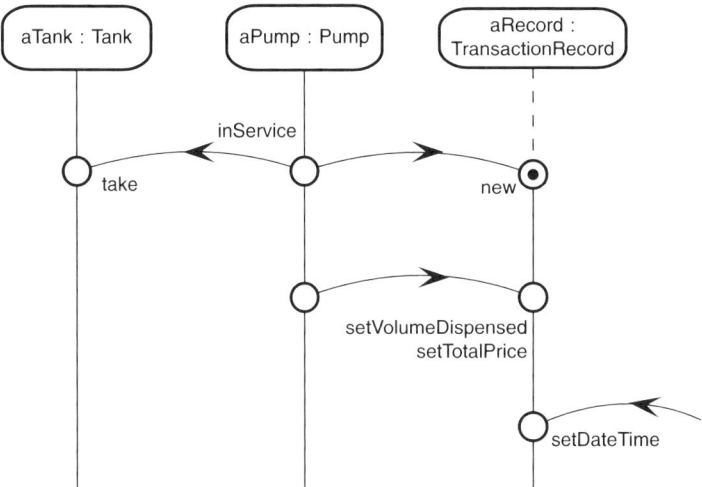

Figure 2.10 Transaction

independent. One of the reasons for modelling dynamics here is to find out how the different kinds of episode interact.

In a *Transaction*, a *Pump* takes a quantity of petrol from its *Tank*, a *TransactionRecord* is created and the transaction details are recorded; the *Pump* must be *inService*—that is, its *Tank* must have petrol (Figure 2.10). The *constituent sets* a *Tank*, *aPump* and *aRecord*, drawn as round-cornered rectangles, are the *stereotypical objects* that take part in a *Transaction* behaviour. The type of each constituent set is given after the colon; these types are the classes *Tank*, *Pump* and *TransactionRecord* shown earlier in the static models. In this case, the behaviour involves only single instances of *Pump*, *Tank* and *TransactionRecord*, but constituent sets can, in general, contain more than one instance.

The vertical lines in the diagram are *timelines*. These are drawn with time progressing down the page. Timelines are used to show *temporal dependency*: ordering of behaviours in time. The circular blobs indicate that a constituent set *participates* in some bit of the behaviour. Curved arrows indicate an *interact event*: a source participant invoking some operation on a target participant. In this example, *aPump* interacts with *aTank*, by invoking the *take* operation. At the same time (more or less), the pump creates *aRecord* by invoking the special *new* operation. The dotted portion of the timeline for *aRecord*, and the blob-with-dot, indicate that this is the first thing that happens to *aRecord*.

After *aRecord* has been created, and the *take* operation has finished, the volume of petrol taken and the total price are recorded. Some time later, the *dateTime* is recorded (by *Payment*, although we choose not to show this here).

When a tank runs out of petrol (the *currentContents* reaches *minContents*), the connected pumps are taken out of service (Figure 2.11). In practice, this happens

20   Chapter 2   ORCA by example

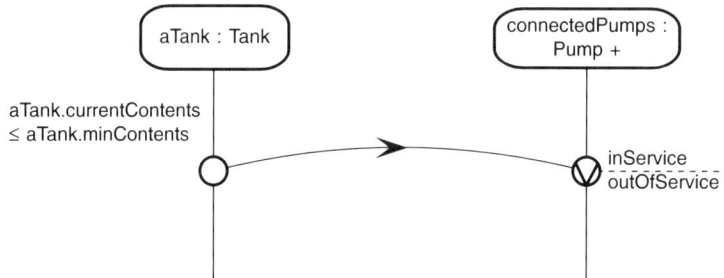

Figure 2.11 Taking Pumps out of service

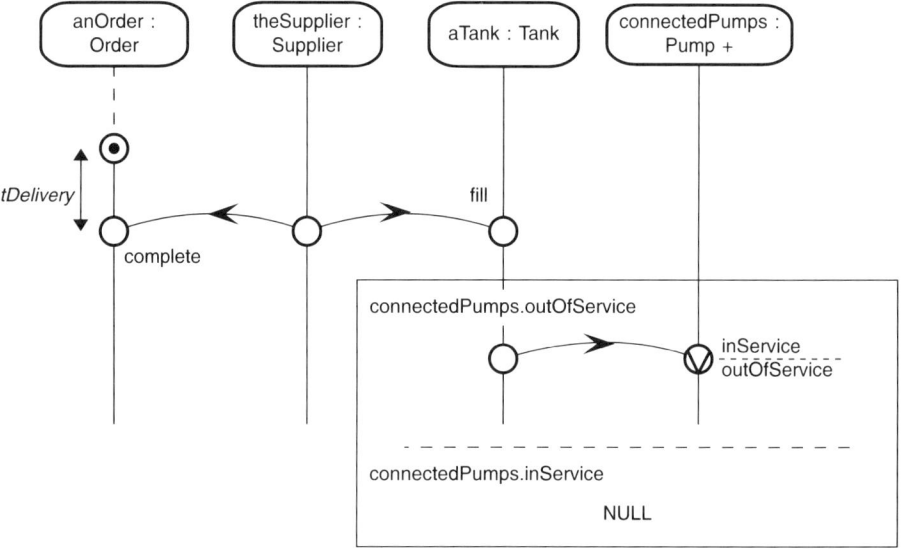

Figure 2.12 Order Delivery

automatically. A *TankEmpty* episode happens only when the condition shown on the left of the diagram is true. The state of the tank causes all the connected pumps to be taken out of service. In this case, the constituent set *connectedPumps* is a set of one or more ('+') *Pumps*. All members of the set change state from *inService* to *outOfService*.

An *OrderDelivery* behaviour has two stages, as its name suggests (Figure 2.12). Some time after *anOrder* is created, it is completed when *theSupplier* fills the appropriate tank, *aTank*. Filling *aTank* puts the *connectedPumps* back in service, if they had previously been taken out of service due to an empty tank. Notice that the model does not say what creates *Orders*; this is an issue to which we shall return.

We are potentially interested in the elapsed time between *anOrder* being created and its completion; this is labelled *tDelivery*. The conditional bit of behaviour that puts pumps back in service is shown in a box, separated from the NULL alternative by a horizontal line. Thus, if the *connectedPumps* are *outOfService*, their state will be changed to *inService*; if the *connectedPumps* are already *inService*, nothing needs to happen.

anOrder is created if and when certain conditions hold. Determining when the appropriate conditions hold is one of the principal tasks of the proposed IT system.

In this model, the three kinds of episode are linked directly by *Tank*s and indirectly by *TransactionRecord*s. In the first case, *Transaction*s take petrol from *Tank*s; in the absence of deliveries, this will ultimately cause a *TankEmpty* behaviour; deliveries counter the depletion process by filling the *Tank*s. In the second case, the information accumulated from the *TransactionRecord*s allows orders to be generated at appropriate times.

2.5.3 Analysis issues

The analysis so far has raised several issues that need to be addressed before we proceed further.

The proposed solution assumes that the typical delivery time, *tDelivery*, is small relative to the time that a tank-full of petrol typically lasts. This allows the ordering behaviour to accumulate information from some initial portion of a fill–depletion cycle, and then generate an order, in the expectation of delivery before the end of the fill–depletion cycle.

If this assumption does not hold, then *OrderDelivery* episodes need to be overlapped; order $n + 1$ needs to be placed before the delivery for order n has been received. In this case, orders would need to be placed on a regular basis, rather than generating them for individual cycles. Monitoring of consumption could be done to adjust the frequency of ordering, rather than the times of individual orders.

For our proposed solution to work well, the time between placing an order and receiving the delivery, *tDelivery*, should have low variability. If this is not the case, the ordering behaviour will need to allow a large margin for error in deciding when to place an order (since it will need to assume the worst case delivery time). One possibility is for delivery times to be monitored, and taken into account in deciding when to place orders. As it happens, a completed *Order* already holds the necessary information, in the form of its *creationDate* and *deliveryDate* attributes.

The ordering behaviour will thus use two bodies of historical information: information about consumption of petrol from the *TransactionRecord*s, and information about delivery times from completed *Order*s. Do we need to take into account any other bodies of information? For example, it is probable that consumption

22 Chapter 2 ORCA by example

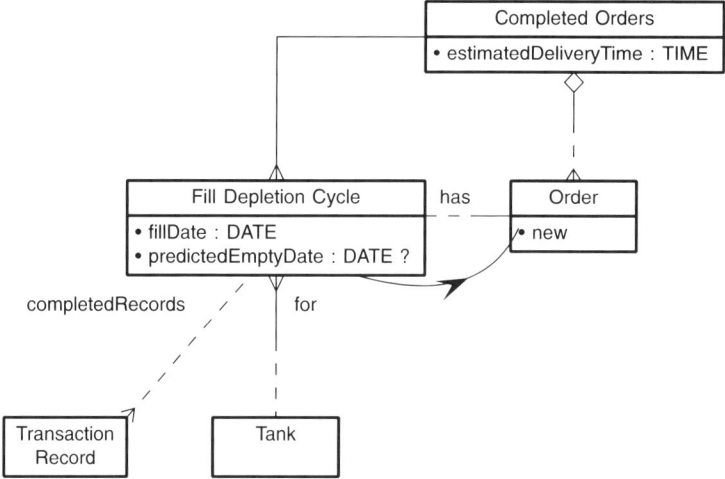

Figure 2.13 The fill–depletion cycle

would increase following price cuts (relative to the prices of neighbouring petrol stations). Such changes in consumption could be anticipated, given notice of the price changes. We might thus want the ordering behaviour to have access to a body of 'plan' information, documenting future events and their predicted effects.

2.5.4 A statement of requirements

The analysis process outlined above has led us to a solution of the problems with the petrol station—essentially, a change in the way that ordering is done. This altered behaviour is to be supported by a new IT system, which will largely automate the process. The models developed in our analysis allow us to give a usefully precise statement of requirements for this IT system.

Basically, the IT system is to accumulate information about petrol consumption from the *TransactionRecord*s, which need to be made available after completion by the payment handling behaviour. The IT system is also to accumulate information about delivery times from completed *Order*s. For each *Tank*, there will be a current 'fill–depletion cycle'. The IT system needs to monitor the consumption within each current cycle, and periodically predict the date when the *Tank* will run out of petrol. A new *Order* needs to be created when the time until the predicted empty-date is close to the estimated delivery time (we need to define precisely what we mean by 'close').

We can now extend our model to make the statement of requirements more precise (Figure 2.13). The start of a *FillDepletionCycle* (*FDC*) is recorded by its *fillDate*. An *FDC* must be associated with a *Tank*; a *Tank* is associated with a succession of *FDC*s. An *FDC* is also associated with a set of *TransactionRecord*s;

2.5 *Specifying a solution* 23

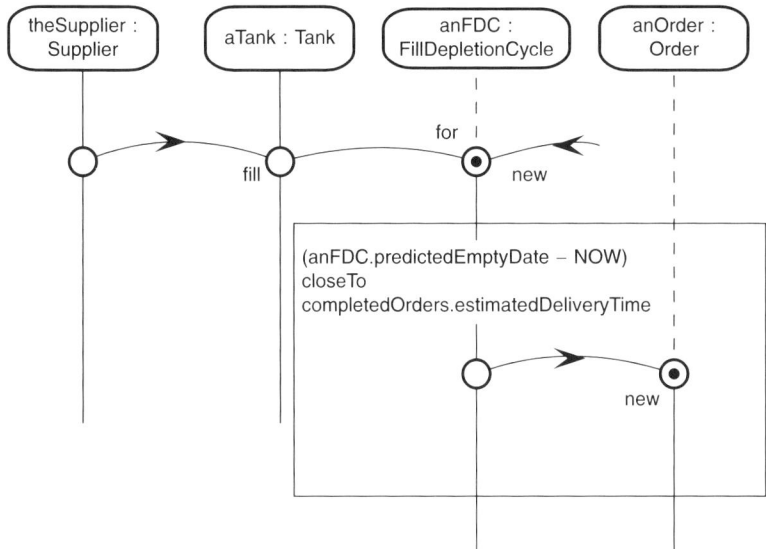

Figure 2.14 Fill–depletion cycle dynamics

this set contains all the records for the associated *Tank* that have been completed since the *fillDate*. The volume dispensed so far in the current cycle can be derived from the *volumeDispensed* attribute of the *TransactionRecord*s, and used to predict the *predictedTankEmpty* date.

A *FillDepletionCycle* is also associated with the collection of *CompletedOrders*; the diamond symbol in the diagram indicates that *CompletedOrders* collects instances of *Order*. The *estimatedDeliveryTime* is an attribute derived from the date information in all the completed *Orders*—for example, it might be the maximum or the median value. At an appropriate time, an *FDC* creates an *Order* (indicated by the curved arrow to the *new* operation).

The dynamic view of this is shown in Figure 2.14. When a *Tank* is filled, a new *FDC* is created (*new*) for that *Tank*—the curved line joining the *aTank* and *anFDC* timelines indicates that they become *associated*. The *fillDate* is recorded by the *new* operation. Some time later, when the 'trigger condition' holds, the *FDC* creates an *Order*. Notice how this behaviour links the previously modelled *Transaction* and *OrderDelivery* behaviours, and shows what it is that creates *Orders*.

If this were a real analysis, there would of course be much more detail and supporting information in the statement of requirements.

2.6 Change and development

Having produced a sufficiently precise statement of requirements, we have reached the end-point of the analysis process. The client now needs to decide whether to proceed with the specified changes to the business and with development of the proposed IT system.

An important issue for the design of the new IT system is how the basic and derived information is to be stored and accessed. For example, the payment handling behaviour writes completed *TransactionRecord*s to some data store, which is also accessed by the ordering behaviour. Since the ordering behaviour accumulates information from *TransactionRecords* as they are produced, storing the records in sequential files may be adequate.

One of the relevant factors in designing information storage is the quantity and complexity of the information to be handled. This is something that should be addressed during the analysis activity. For example, it will be useful to know how many transactions typically take place within a fill–depletion cycle.

As developers, we also need to determine the tasks undertaken by or with the IT system, and when these are performed. A simple option would be to run a batch process at the start of each day. This process would:

- collate information from the previous day's *TransactionRecord*s
- for each current *FillDepletionCycle* (one for each *Tank*), calculate the *predictedEmptyDate*, and determine whether an *Order* needs to be placed
- generate *Order*s, both as internal information, and as concrete outputs (for example, paper forms)

Updating the *estimatedDeliveryTime* would probably be done on a less frequent basis, depending on how often deliveries occur—perhaps once a week.

We also need to consider user interaction, generation of management reports, whether there needs to be manual over-ride in case the system doesn't work properly, and a variety of other issues.

2.7 Conclusion

This relatively simple example illustrates the process of analysis, from the very earliest stages through to a precise statement of proposals for change and (in this case) requirements for an IT system. The analysis, if done in full, would provide a firm basis for proceeding into IT systems development.

The next chapter summarises the key ideas of ORCA, and presents them in the context of this example.

Chapter 3

ORCA's key ideas

3.1 An analysis process needs to be tailorable

Different analysis situations require different analysis processes. In some situations, particular aspects of the analysis process need to be emphasised, while other aspects are de-emphasised or omitted altogether. It is not necessary, or even desirable, to use all of ORCA on every analysis! For example, some aspects might be entirely absent where the client's world is developing into new areas, or is responding to external change. One of an analyst's first tasks is to assess the nature of the given analysis situation, and determine the kind of analysis process that is appropriate.

An analyst therefore needs to respond to different situations by *tailoring* the analysis process. Part II illustrates ORCA's *Basic Process* by working through an example analysis in detail; Part IV presents a variety of process tailorings, and discusses their applicability.

This approach to analysis is reflected in the ORCA method itself. Although we refer to a 'method', this does not denote a set of procedures to be worked through mechanically. Rather, ORCA offers a set of ideas and techniques that we hope is both coherent and practical.

3.2 A system has purposes as well as behaviour

A system is not just a 'naturally occurring' combination of objects and interactions—we assume that it is the subject of various *purposes*. For the petrol station's owner, the purpose of the petrol station is to make money; for the customers, it is to supply petrol to them, when they want it, at a reasonable price; for the attendant, to provide employment; for the supplier, to buy its petrol. These purposes are what gives a system its *raison d'être*.

The various purposes of a system may or may not be compatible with each other. This is something that the analysis process needs to discover, since incompatible purposes may be a root cause of observed problems.

The analysis process also needs to investigate the degree to which the behaviour of a system satisfies its purposes. In the case of the petrol station, the system is deficient in the degree to which it satisfies its purpose with respect to the customers (providing petrol, ...). As a consequence, via the problem of falling sales, it is deficient in the degree to which it satisfies its purpose with respect to the petrol station owner (make money).

It follows from this view that analysis is concerned not only with *behavioural modelling*; *purposive modelling* is also needed, in order to understand what a system is supposed to be doing. ORCA provides two modelling languages to support this dual approach.

3.3 A system is a network of co-operating roles

The purposive view of a system is as a network of co-operation relationships between *roles* with regard to the provision of *services*. These *co-operations* exist (or should exist) in order to achieve the various purposes within a system.

A co-operation between two roles exists when one role guarantees to provide some service to the other role, and the second role relies on the provision of that service. In general, multiple guarantees may combine to satisfy a single reliance. ORCA's purposive modelling language allows co-operations between roles to be stated explicitly. The models in Sections 2.3 and 2.4 show how a system can be described by a network of co-operations.

In a properly functioning system, each co-operation is realised by some behaviour. Objects can *interact* either directly, or via 'shared' objects. Objects can become mutually *associated*, and conversely dissociated, at different times. ORCA's behavioural modelling language allows patterns of interaction and association between objects to be described, using the framework construct.

In the purposive view of a system, each role can have its own local view of the world, in terms of that role's guarantees and reliances, and the co-operations in which it is involved (either as guarantor or relier). The analysis process may thus need to check the compatibility of multiple *viewpoints*, and reconcile incompatibilities where these exist. ORCA's purposive modelling language supports the description multiple viewpoints. Section 2.3 gives a local description of Customers, but in that simple example we do not have the problem of incompatible viewpoints. We might have had the situation shown in Figure 3.1. There are three problems, shown in the diagram by using dashed lines for the co-operations. Customers want cheap petrol, and will presumably go elsewhere to find it, whereas the petrol station provides petrol at a price that is determined by factors other than customer demand (such as OPEC agreements and excise duties). The petrol station seeks

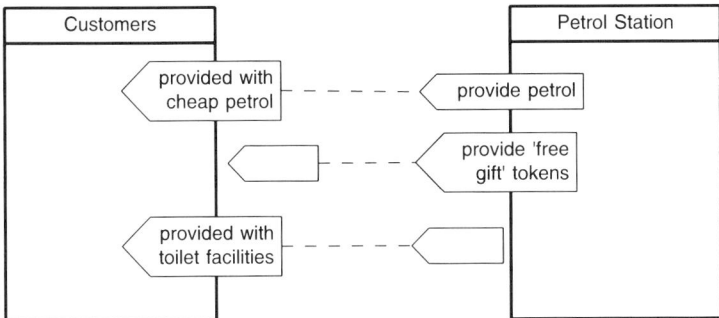

Figure 3.1 Problematic co-operations

to gain customers from its competitors by offering tokens for attractive 'free gifts'. There is a problem here because customers don't actually want these—they would rather have cheap petrol. Customers also want the petrol station to provide toilet facilities, but the petrol station seems unaware of this.

The petrol station's view that it should provide gift tokens is an *intrinsic* description (what the petrol station believes about itself). The petrol station's belief that customers want gift tokens is an *extrinsic* description of Customers. In Figure 3.1, this is shown by the guarantee box adjacent to (not overlapping) the Customers box. Similarly, the customers' belief that the petrol station should provide toilet facilities is an extrinsic description of Petrol Station. In ORCA's purposive modelling language, extrinsic guarantees or reliances are shown only if there is no matching intrinsic description. For simplicity, Chapter 2 assumes that extrinsic and intrinsic descriptions match, and that all co-operations are valid (indicated by a solid connecting line).

3.4 A system may exhibit pathologies

The aim of analysis is to determine the requirements for change to the world. The old adage 'if it ain't broke, don't fix it' suggests that proposals for change should arise from an analysis of what is wrong with the world. Using a medical metaphor, we refer to the latter as a *pathology*; the proposals for change are referred to as the *prescription*.

A pathology might reflect internal defects, such as inappropriate organisation or incompatibilities between components. Alternatively, the previously viable world might be inadequate in the face of external changes, or might have inherent limitations that are preventing desired change. With the petrol station, the problems are internal—ordering is not working effectively.

ORCA's dual view of the world, in terms of both purpose and behaviour, allows us to characterise the following different kinds of pathology:

- Disagreements between roles about what they should be doing: that is, faulty co-operations. In ORCA purposive models, these show up as reliances that are not met by guarantees, or unmatched extrinsic descriptions.
- Faulty realisation of co-operations by behaviour: the services that should be provided are not provided adequately.
- Faulty or inefficient interactions, due to behavioural incompatibilities, lack of coordination or insufficient resources.

A prescription proposes changes to the world in order to remedy a pathology. These changes might involve any of the following:

- reorganisation
- reallocating responsibilities between components
- changes in manual procedures
- improvements in communications
- increases in resources
- staff training
- capital purchase of equipment or facilities
- computer support

With the petrol station, the prescription is basically for a change of procedures together with computer support, but capital purchase and staff training might also be involved. The examples in Part IV illustrate different prescriptions.

For a given pathology, a number of alternative prescriptions can be produced. These may then be assessed against relevant criteria, such as the time and cost of implementation, running and maintenance costs, or level of risk involved. With the petrol station, we might have decided that it would be simpler and cheaper to install additional storage tanks.

3.5 Models can describe the Old World, or a New World

Given that a prescription is a proposal for change to the world, we potentially have two kinds of model: of an *Old World* that exhibits a pathology, and of a *New World* that exists after implementing a prescription. In Sections 2.3 and 2.4 we have Old World and New World purposive models of the petrol station. These indicate the need for an additional co-operation between Payment and Ordering.

The behavioural model of tanks, pumps, transaction records and orders is common to both Old and New Worlds. This is because our proposed solution is essentially to do with decision making and scheduling (creating orders at more appropriate times), rather than the introduction of new kinds of object. However, the 'intelligent' fill–depletion cycles belong only to our New World (and implementing the computational aspects of these objects is a task for the software designers).

Just as we may have alternative prescriptions, so we may have alternative New Worlds. In some cases, we might want to have an explicit description of an 'ideal' world, regardless of the practicalities of realising this. Alternative 'pragmatic' New Worlds could then be assessed against this ideal.

Although the basic ORCA process covers the four related ideas of Old World–Pathology–Prescription–New World, these are not equally applicable in all situations. Where the client's world is developing into new areas, for example, there may be nothing useful to say about the Old World. Where external changes are forcing re-evaluation of a previously viable system, the Old World will not be pathological in itself but only in the context of the external changes.

3.6 A system is embedded in an environment

The 'real world', which analysis seeks to describe, is a continuum—there are no hard boundaries between different areas. In contrast, a model of a system is necessarily 'closed': it is a statement of what is relevant for the purposes of analysis. Things not stated explicitly are by implication irrelevant.

What this means in practice is that analysis should investigate the 'environment' of a system, in order to determine what is actually relevant. From a given 'core' system, chains of purposive and behavioural relationships can be followed outwards. At some stage, a decision must be made that the 'system boundary' has been reached, and that everything outside this boundary is viewed as irrelevant (for the particular analysis).

In Chapter 2, the petrol station itself is the 'core' system. However, the problems that the analysis is investigating require customers and petrol suppliers to be included in the system. We could have cast the net wider still, and looked at how suppliers get petrol from producers, or how national taxation authorities affect the price of petrol. There is thus no *a priori* boundary to a system: it must be determined by analysis, for a particular analysis situation.

The 'core' system from which the investigations start typically corresponds to that part of the world over which the client has direct control. In our example analysis, this is the petrol station as a commercial unit. Radical changes (including IT developments) are potentially feasible. In contrast, the petrol station's environment contains customers and supplier, which are not under the client's direct control. Changes to the system involving such 'environmental' entities are going to be much harder to implement, although it might be possible for the client to influence them, in various ways. In practice, there may be constraints on what can be changed within the client-controlled world, for political, financial or practical reasons. What is or is not open to change may become apparent only during analysis.

3.7 IT systems are ways of implementing behaviour

The analysis process may need to talk about 'IT systems', either as part of the Old World (existing data processing or control systems, say), or as part of a proposed New World. For the petrol station, we propose a computerised ordering system as part of the New World.

Such IT systems are different in nature from the 'systems' that are the subject of analysis. The latter kind of system is a view of what is essential about some area of the world (in both purposive and behavioural terms). IT systems are a way of implementing some aspects of essential behaviour.

In the case of the petrol station, the proposed New World involves more 'intelligent' ordering behaviour, using information about customer transactions. This new behaviour needs to be implemented using some combination of human, physical and computational resources. The idea of an 'IT system' is an IT developer's view of this: 'users'; input/output devices, communications links; software data and functionality.

For the purposes of analysis, we are concerned with the capabilities of existing or proposed IT systems, rather than their actual design (although it is difficult to divorce the two entirely). As with all prescriptions for system change, specifications for IT developments need to be practicable, given some set of development resources and constraints. The aim of analysis should be to define a solution that is both effective and practicable, while not unduly restricting the ways in which the solution can be implemented.

3.8 Models can express different levels of abstraction

Both purposive models and behavioural models can express different levels of *abstraction*. By this we mean whether a modelled entity is viewed as a single unitary entity (a role or an object) or as a composite of multiple simpler entities (a cluster of subroles, or a population of subobjects).

Within a purposive model, a role can be modelled as a cluster of roles; each component (sub)role can in turn be modelled as a cluster of (subsub)roles, and so on. A service guarantee offered by a role can be *delegated* to guarantees of one or more of its component roles. Conversely, the need of a role to rely on a service can be *promoted* to its enclosing role. Examining how guarantees and reliances are propagated between roles at different levels is an important part of purposive modelling in ORCA.

This is illustrated by the petrol station example, in which services involving the petrol station as a whole are delegated to (or promoted from) its component roles. For example, Petrol Station guarantees to Customers to 'provide petrol', and this is delegated to Petrol Dispensing (Figure 2.4).

3.8 *Models can express different levels of abstraction* 31

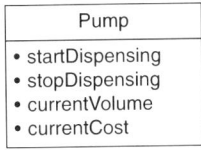

Figure 3.2 Pump as a class

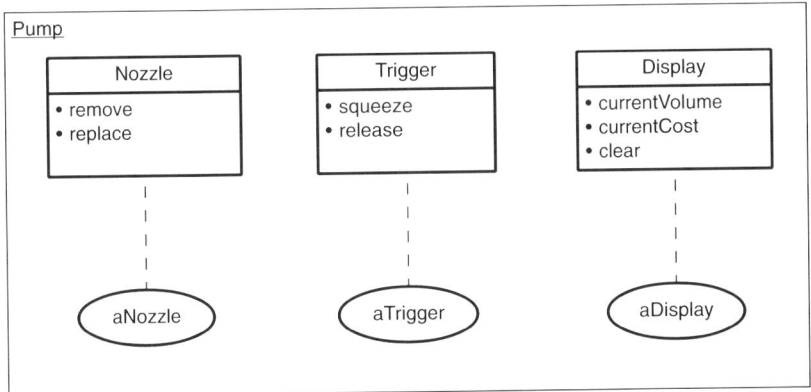

Figure 3.3 Pump as a framework

Within a behavioural model, unitary objects are treated as instances of *classes*, where the class of an object defines its *features* (attributes and operations). In the second case, objects are treated as populations of subobjects, described by *frameworks*, which define organisational structure and patterns of dynamic behaviour.

Suppose that we are interested in a customer's interaction with a pump. We could characterise pump objects by the class *Pump* (Figure 3.2). *startDispensing* and *stopDispensing* are operations performed by a *Customer* on a *Pump*. *currentVolume* and *currentCost* are visible attributes, shown on the pump's display.

Alternatively, we could treat a pump as composed of its parts (Figure 3.3). The diagram says that a *Pump* consists of one *Nozzle*, one *Trigger*, and one *Display*—there would in practice be other parts as well.

The features defined by the *Pump* class can be mapped to the constituents and behaviours defined by the *Pump* framework. In this case, the *startDispensing* operation on *Pump* maps on to a combination of operations: removing the nozzle from its holster, then waiting for the display to be cleared, then depressing the trigger. A similar situation holds for *stopDispensing*. The *currentVolume* and *currentCost* attributes are mapped to equivalent attributes of *Display*.

Within analysis, models are used to investigate a system, and to provide some definitive description of it. In the first case, it is important that an analyst has the conceptual and notational tools to explore different levels of abstraction. ORCA attempts to provide these tools. In the second case, it is important for an analyst to find an appropriate level of abstraction at which to describe the system. A sufficient, but not excessive, amount of detail needs to be provided.

The petrol station example is too simple for this dimension of analysis to be very evident, but the larger case study in Part II illustrates it in more detail.

3.9 Overview of ORCA

In summary, the key ideas of ORCA are:
- An analysis process needs to be tailorable.
- A system has purposes as well as behaviour.
- A system is a network of co-operating roles.
- A system may exhibit pathologies.
- Models can describe the Old World, or a New World.
- A system is embedded in an environment.
- IT systems are ways of implementing behaviour.
- Models can express different levels of abstraction.

The second and third items in this list concern the conceptual basis of ORCA. The key concept is that of a *service*: a system is seen as a set of interrelated services. ORCA offers two views of services (and thus of systems): purposive and behavioural. This is illustrated in Figure 3.4. This dual view of systems is explored more fully in Part III.

3.9 Overview of ORCA   33

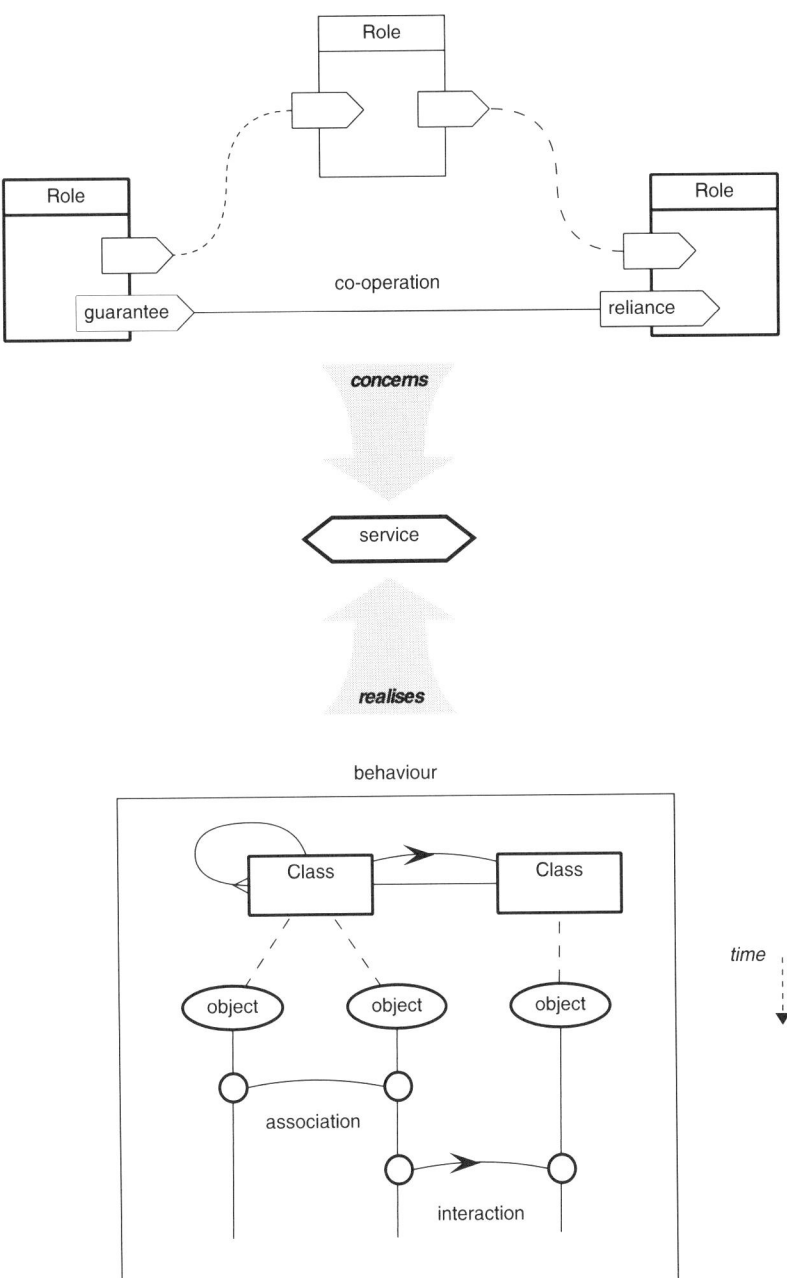

Figure 3.4 ORCA conceptual framework

Part II

The Basic Process

Chapter 4
Introduction to the case study

4.1 Introduction to Part II

In this part, we work through a substantial, realistic case study. This case study is used to introduce and illustrate the main aspects of ORCA. Details of the modelling languages are explained in the context of the example material, while general discussion of modelling concepts is left until Part III.

The chapters in this part take the case study through the whole analysis process. The reader might wonder why such an extensive example is required. There are two reasons. Firstly, analysis typically deals with complex situations. This is precisely when the assistance of an analysis method is needed. If this book were to deal only with 'toy' examples, it would appear as if we were using the proverbial sledgehammer to crack a nut. Secondly, it would be difficult to illustrate the time dimension of the analysis process using a small example. Analysis proceeds by integrating different partial views of a system into a coherent picture, and constructing a specification and rationale for change. This aspect of the process would not be illustrated by simply presenting a finished 'solution'.

The subject matter of the case study—a weaving factory—is deliberately unfamiliar, and comes with its own (somewhat bizarre) terminology. As with complexity, this reflects the real world situations with which analysis typically deals. The method would not be sufficiently exercised if we 'analysed' a familiar, unproblematic situation.

The analysis case study, as written up for this book, is necessarily a rationalised and tidied-up version of what was produced in actually doing the case study. Nevertheless, the chapters in this part do provide a fair illustration of how ORCA can be used in practice.

This introductory chapter serves two purposes. Firstly, it provides a preview of the ORCA Basic Process, which is used for the case study. (Variants of the Basic Process are illustrated in Part IV.) Secondly, it provides a brief description of the analysis situation addressed by the case study, in lieu of real information gathering.

4.2 Overview of the ORCA Basic Process

We use a basic version of ORCA's analysis process for the case study described in this part. This Basic Process covers most of the products and activities that make up ORCA, and does so in a fairly straightforward sequence. In practice, no single process is suitable for all analysis projects, so it is likely that the Basic Process will always be tailored to some extent. Examples of such tailorings are described in Part IV.

All ORCA projects, whether using the Basic Process or a tailored one, begin with the following activities:

Preliminary Analysis; Information Gathering

A problem statement, whether a formal proposal or an informal musing, can express discontent with a current situation, or a desire for things to be different. Such a statement forms the starting point for an analysis project. Some consideration of the problem statement is needed in order to understand whether anything useful may be done at all.

If the project is to proceed, a common understanding of the scope and aims of the project is necessary for the client and analyst. The project's scope should give an indication of which areas are to be investigated, and how thoroughly. The project's aims are likely to be given in relation to development objectives—removing a problem or building something new. An agreement achieved in this activity can form the basis for a formal (contractual) definition of the project.

Process Design

To be effective, the process to be used by the analysis project has to be designed with regard to the expected characteristics of the project—the scope and aims of the project, and its participants. A process describes a bundle of activities and their products, and a plan for the execution of this process. The plan covers resource management and monitoring throughout the project. The Basic Process covers most of ORCA's activities and products.

The Basic Process continues as follows:

Modelling Old World Purpose and Behaviour

Faced with a complex or novel situation, the analyst may feel the need to model the way things currently are, in order to understand the situation better. Two aspects of the world can be modelled: purposes, and the behaviour supporting those purposes. Only as much modelling should be done as is needed to clarify the situation, or to identify problems with the current situation, or to identify parts of the world that are affected by development objectives.

Determining Pathology; Prescribing Change

The Old World model provides a basis for determining system pathology: the identification of fundamental problems with the current situation, particularly with regard to any new demands imposed by the development objectives. The consequences of these problems can be traced throughout the system, their seriousness assessed, and courses of action prescribed for resolving the problems. A prescription can encompass changes to the existing system and the development of new system components.

Modelling New World Purpose and Behaviour

A set of prescriptions that are compatible with each other can be considered as one bundled course of action. Several alternative bundles might be evaluated with regard to how well they achieve the development objectives, and what other benefits and costs are associated with them. One of these bundles may be chosen as the one to be implemented.

Models of the bits of the New (post-development) World that the analyst is interested in are built. These can again be in terms of purposes and the behaviour supporting those purposes. These models form the New World specification, and, unless the proposed changes are extremely radical, will be derived in part from the Old World models. The specification is the main tangible 'product' of the Basic Process, although the confidence in the chosen course of action given by a thorough understanding of the Old World, and of the problems and possible changes that could be made, is also important.

Development and Transition

After these activities comes system development, which involves implementing the New World specification, and so making changes to the real world. This activity is shaped both by the characteristics of the project as identified in the initial activities, and by the nature of the prescriptions chosen. It is even more unlikely that one process covers all eventualities here, and so this activity must begin with choosing a process that is effective in allowing the chosen New World to be built. The activity continues as detailed in the chosen process.

This activity is strictly outside the scope of the Basic Process, being the activity during which the Basic Process stops being used, and something else begins to be used. For a successful project, the two have to be compatible. Most directly, the New World specification must be appropriate as raw material for the chosen process—or the process that is chosen must be appropriate for the New World

40   Chapter 4 Introduction to the case study

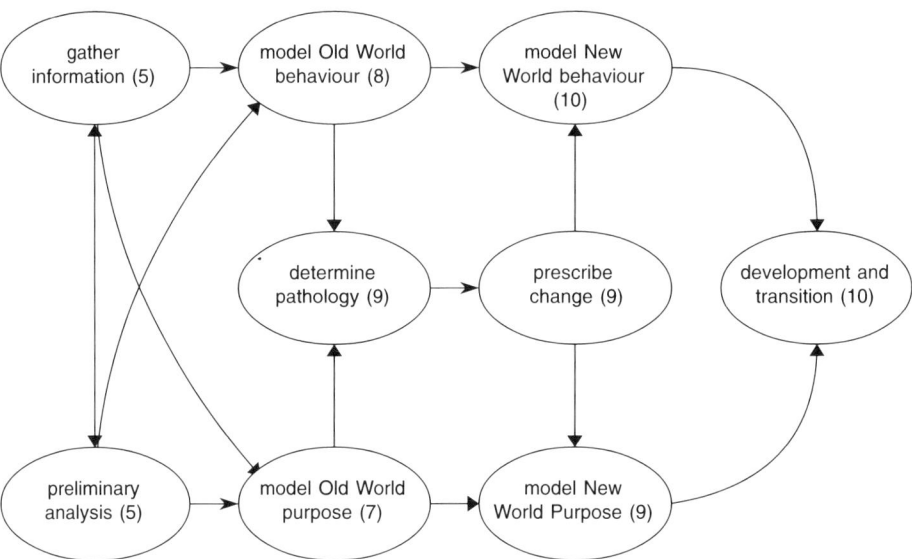

Figure 4.1 ORCA's Basic Process. The bubbles represent activities, the arrows represent dependencies. The numbers point to chapters in this part where the activity is described

specification that has been produced. More is given on this post-ORCA activity in Chapter 22.

The Basic Process is summarised in Figure 4.1. An iconic 'road map' form is used to head each remaining chapter in this part, to indicate which activities are being described there. Also, variants of the diagram are used in Part IV to indicate which parts of the process the tailorings affect.

4.3   Brief description of the case study

The case study described in this part concerns a (real) small weaving company. Since this is, unusually for a weaving company, based nowhere near Manchester, we have dubbed it NIMWeC: the 'Not In Manchester Weaving Company'.

Below, we give a brief description of the analysis situation for NIMWeC. This provides some basic information on the business and the manufacturing process, which the analysis will attempt to clarify. It also provides an initial statement of the client's current concerns and aims for the future, which the analysis will address. A glossary defines some of the weaving-specific terminology.

This description is provided in lieu of actual information gathering: talking to the client and the employees, observing the manufacturing process, studying paperwork, and so on. This is a necessary ploy for the purposes of this book, since it would be difficult and tedious to mimic actual information gathering (for example, by giving transcripts of hypothetical interviews). The description below is *not* an initial 'problem statement' given to us as analysts. The case where a documentary 'problem statement' is the only source of information is dealt with in Chapter 20.

4.3.1 NIMWeC

NIMWeC is a narrow fabric weaving factory. There are two main products, which are kinds of woven label:
- name tapes, sold mainly to parents for sewing into their children's school clothing
- shirt size labels, sold to shirt manufacturers

Name tapes make the money, but are seasonal. Shirt size labels are 'loom fodder' to keep the factory and staff employed when the name tape business is slack.

4.3.2 How name tapes are ordered and processed

Orders for name tapes come from two sources:
- Books of 50 order forms for name tapes are sold to school clothing shops. A customer buys an order form from the shop, fills it in, and the shop sends it to NIMWeC.
- Adverts containing an order form are placed in magazines. A customer fills in the order form and sends it, along with payment, to NIMWeC.

A correctly filled-in order form contains five items of information:
- the customer's address
- the name to be woven—there are no limits on length
- the quantity, in multiples of 72 (a *repeat*)
- the letter colour—red or blue
- the style required—large or small

Orders are batched into numbered groups of about 30 repeats, each batch composed of orders that as far as possible all have the same style and colour. These batches are sent to the factory floor.

Sometimes an urgent order subverts the batching process by being sent straight to the looms. This is done, for example, when an order has been woven incorrectly, and needs to be rewoven.

There is a busy period towards the end of the summer, as a consequence of the school year beginning in September. All orders are woven within 14 days—NIMWeC's big competitive advantage.

4.3.3 How labels are woven

The current factory is based around two looms and their control systems. Some characteristics of the loom are:

- Each loom has one operator.
- Each loom weaves 72 pieces at a time (hence name tapes are ordered in multiples of 72).
- Each loom runs at about 100 picks/minute.
- Each loom has a 'warp end breakage detector', which stops the loom should a warp end break. A warp end breaks every half hour or so during weaving.
- Each loom has two electromagnetic jacquard machines, which lift or drop warp threads to allow a shuttle to be passed through.
- Each loom has an electromagnetic lifting box, which selects whether a brocade or a ground shuttle should be used on the next pick.
- Each loom has a cam arrangement whereby the jacquard can detect what stage of weaving the loom is at, and so whether to lift or drop threads.
- Each pair of jacquards is connected to a control box, which tells the jacquards which warp ends to lift or drop for each pick.

Only one jacquard machine would be needed for each loom if the height of the loom were increased by a couple of feet and some new harnesses fitted.

A white ground warp is used. A loom weaves only one colour at a time (so all the brocade shuttles have the same colour in them). Changing the colour of a loom takes about an hour.

The warps on the looms last about a year. The warps must not run out during the busy season, as it takes about two weeks to re-warp a loom.

Each of the looms is controlled by a control box, the front of which has four patch panels, each of which allows the operator to set up one name to be woven. Each of the control boxes controls one font: style large (loom 1) and style small (loom 2).

4.3.4 Name tape dispatch

Special marker patterns are woven into the ribbon: a mark to show the beginning of each batch; a fold mark to show where to fold the ribbon when sewing it into clothing; a mark at the boundary between different names and the boundary between repeats of the same name. A tie-mark is woven along with the fold mark;

this ties the brocade thread to the back of the ribbon to stop it pulling out of the cloth when the ribbon is cut up.

The ribbons that come off the looms are cut up into individual packs of names and put in plastic bags, one batch to a bag. They are then returned to the orders office.

4.3.5 Problems and worries

- Patch panel: the operator can be at the most three names ahead of the loom. Setting the correct plugs in the correct holes sufficiently ahead of the loom is a major problem requiring significant staff effort. Customers do return incorrectly woven orders to NIMWeC.
- Control box: these machines are very expensive and very unreliable, and so no more should be bought.
- Orders are sometimes sent to the wrong address.
- As the majority of the factory's work is done during the summer holidays it is essential that everything works reliably during this period.

4.3.6 The client's idea of a desirable state of affairs

The broad requirements are:
- double production of name tapes (there is sufficient market demand)
- more statistics about the process (this should allow better prediction of future activity)
- fewer incorrectly woven name tapes
- the ability to expand into the Israeli and Swedish markets

4.3.7 Weaving glossary

jacquard: the mechanism responsible for forming the correct **sheds** for the letter being woven. NIMWeC's looms have an electromagnetic arrangement connected to the control box. Each jacquard machine detects where in the loom's cycle it is, so that it knows when:
  - the jacquard electromagnets should be energised
  - the lifting box electromagnets should be energised

lifting box: controls which **shuttle** (ground or brocade) is used on the following **pick**.

narrow fabric loom: weaves ribbons with 64 threads, about 1 cm wide. Each of NIMWeC's looms weaves 72 ribbons at a time.

patch panel: a panel on the control box into which an operator puts pegs, each peg corresponding to a letter of the alphabet, to weave a particular name.

pick: one loom cycle.

piece, ribbon: a strip of cloth woven by a narrow fabric loom.

shed: the pattern of **warp thread**s picked up or dropped down at each **pick**.

shuttle: used to hold a bobbin of **weft thread** and pass it between the separated **warp**s. The ground and brocade wefts are each mounted in a bobbin in a single shuttle; there are two shuttles per ribbon.

warp thread, warp end: the thread that runs in the direction of weaving, along the length of the ribbon.

weft thread: the thread inserted across the direction of weaving. There are two kinds of weft:

> **ground weft:** thread that is the same colour as the warp, making the 'body' of the ribbon.
>
> **brocade weft:** red or blue brocade, inserted to make the letters. When the brocade is inserted the loom does not move the ribbon on, in order to avoid unsightly gaps in the ground ribbon.

Chapter 5
Preliminary Analysis

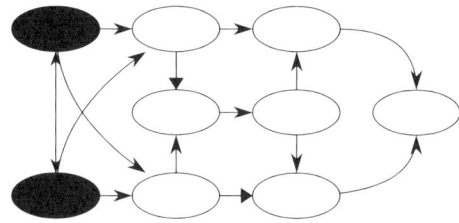

5.1 Introduction

This chapter is about the preliminary activities that need to be performed in order to set the scene for the main analysis activities. This 'preliminary analysis' involves

- gaining an initial understanding of the client's world
- establishing the analyst's terms of reference
- identifying the sources of information
- defining the contractual task

The rest of this chapter looks at these activities in more detail.

5.2 Initial understanding of the client's world

It is important for an analyst to get an initial understanding of the situation to be analysed. Although this understanding might well change in the course of analysis, it provides necessary orientation during the early stages of the process. Initially, an informal overview of the client's world should be built up in discussions with the client.

A useful technique is to use a Soft Systems Methodology *rich picture* [Checkland 1981], [Patching 1990]. A rich picture is an informal pictorial description of a system, suitable for constructing interactively on paper or whiteboard. This allows a description of the client's world to be captured in a way that allows discussion to take place, and consensus and conflict to be identified. There are no rules for rich pictures—it is up to the writer to determine how to capture information. Hence rich pictures are not intended for other people to read (although they could be 'talked through' by the writer); it is the process of drawing the picture that is important, rather than the end-result.

During initial discussions with the NIMWeC client we draw the rich picture shown in Figure 5.1. This rich picture includes the things that seem most important to the client, such as looms, control boxes, customers, warps, orders, and so on.

5.3 Establish the terms of reference

A project's *terms of reference* establish the framework in which the analyst and client operate. The terms of reference should define the nature and extent of the analysis, the aims of the analysis and subsequent development, and the people to whom the analyst is responsible. It is impossible to give an exhaustive list of the contents of the terms of reference, as they will vary according to the nature of the project.

For the NIMWeC analysis, we consider the following issues:
- drawing the analysis boundary
- determining the political and social context
- identifying the development objectives
- identifying the analysis objectives

5.3.1 Draw analysis boundary

Setting the analysis boundary delimits the areas of the client's world to be investigated. This boundary must encompass all areas that may be subject to change and development.

The initial problem presented by our NIMWeC client is expressed in terms of problems associated with the control boxes for the looms. These are expensive and unreliable, and the client wants to replace them with a new control system in order to be able to expand production by operating more looms. So, the manufacturing process is to be our primary area for investigation.

However, after discussion, we agree with the client that the scope of analysis should be widened to include NIMWeC's interactions with its customers and suppliers. This allows us to investigate the wider consequences of an expansion in production, both with regard to the other parts of NIMWeC, and with regard to the outside bodies that it deals with (Figure 5.2).

This set of concerns corresponds to a high level view of manufacturing, with customers and suppliers, and NIMWeC in the middle.

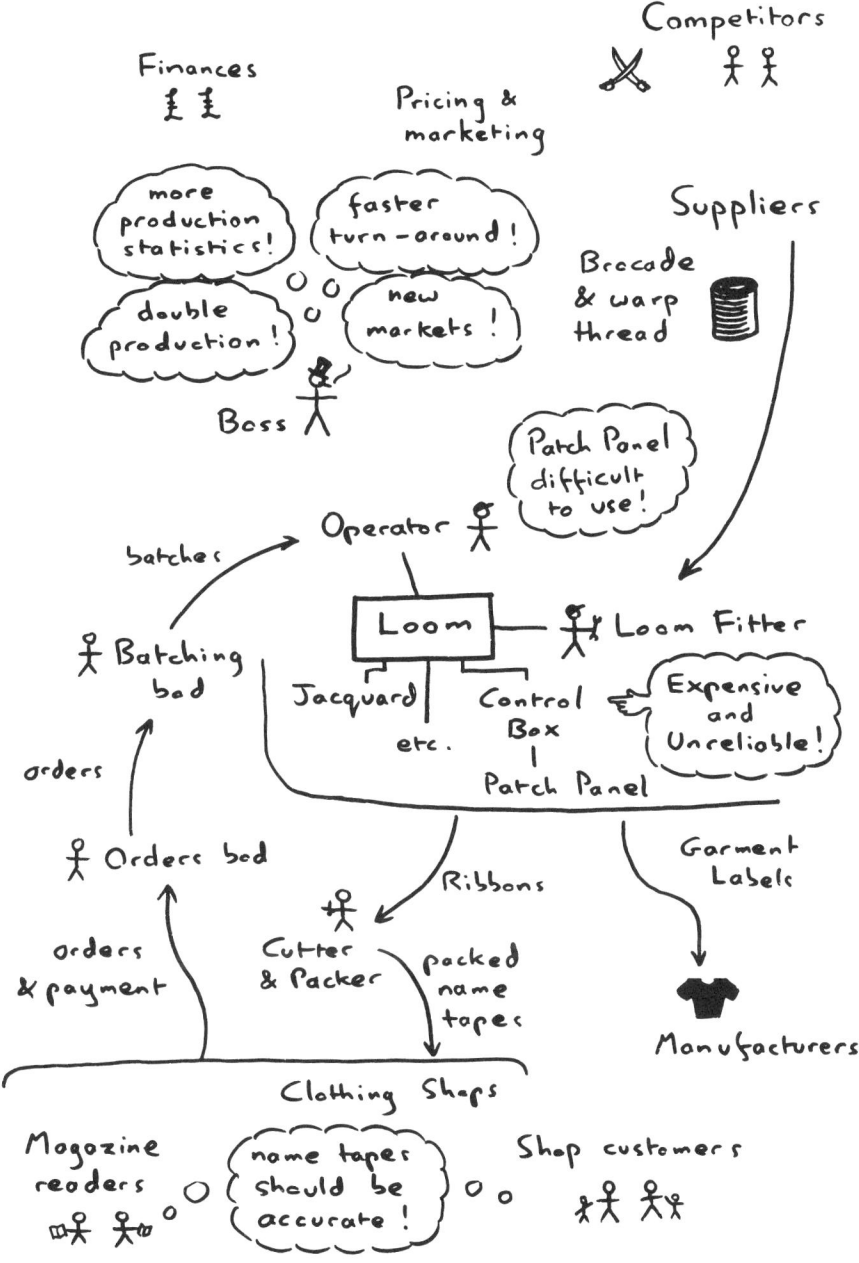

Figure 5.1 NIMWeC rich picture

48   Chapter 5   Preliminary Analysis

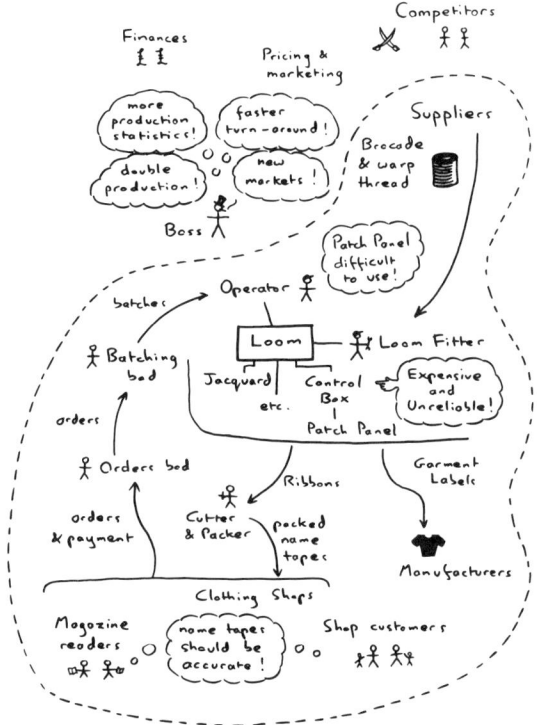

Figure 5.2 NIMWeC analysis boundary

It is important to establish what falls outside the analysis boundary, as well as what falls within it. In this case, the client excludes financial matters, pricing policy and marketing activities. In other words, the analysis is to look at NIMWeC as a manufacturing system, rather than as a commercial system. Also, although the boss (our client) acts as a source of information and objectives, his role within the company is not open to examination. Proposals that NIMWeC be turned into a workers' co-operative will not be welcomed!

5.3.2   Determine the political and social context

An analyst needs to understand the political and social context within which investigations and changes are to take place. The following questions give an idea of what an analyst should find out.

- Who has the authority to implement changes in different areas?
- Who resolves conflict if this occurs?

- Who wants change to happen? Who doesn't?
- Are certain kinds of change going to provoke opposition? If so, why?
- Are particular ways of doing things valued?
- Who ought to be involved in the analysis process?

An analyst also needs to consider the motivation for the analysis project itself. A common scenario is that one faction within an organisation obtains the services of external consultants to reiterate their own views to other factions (perhaps to senior management). Disinterested opinions from an external source are often assumed to carry greater weight in internal disputes ('and if they're this expensive, they must be right'). In such a situation, a particular conclusion may be expected from the analyst. This clearly presents problems if the analysis points to another conclusion.

Formulating a strategy in the light of these investigations will help to steer a course through the often murky political and social waters of a client organisation.

For NIMWeC, we establish that the company is under the ownership of a single person. The owner—our client—manages the company directly and takes all major decisions.

There is a small workforce made up mostly of long-term employees. Some of the staff have valuable skills and many have extensive knowledge of the operation of NIMWeC. The client is adamant that these staff should not be deskilled if at all possible—they should retain a high degree of control over the weaving process.

There is dissatisfaction with the looms' control boxes, and a desire for a more up-to-date working environment. It seems that any prescription is unlikely to be considered unless it addresses these problems. However, the client also feels that the character of the factory (dark, noisy and generally with a feel of the workhouse about it) and many of the traditional aspects of the weaving process should be retained.

5.3.3 Identify development objectives

Clients typically expect the analysis and resulting developments to assist in achieving specific development objectives, and possibly also to achieve more general objectives for the organisation as well. It is important that the analyst understands all these expectations. In particular, the solutions recommended must support both sets of objectives, and not undermine them.

Our client wishes to improve profitability by reducing errors in name tape weaving, by selling twice as many name tapes, and by being better able to predict demand. The strategy that the client is pursuing is to increase production and efficiency, and perhaps to sell into foreign markets (Israeli and Swedish) in the future, if all goes well.

5.3.4 Identify analysis objectives

Analysts and clients must have a common understanding of the expected outcome of the analysis. If the analyst is intending to generate only a high level strategic analysis of the organisation and the client is expecting a detailed specification of new IT systems, then there are going to be severe problems.

In this case, our client requires

- a specification for a new control system for the looms, with a view to increasing and ideally doubling production, and eliminating the error-prone parts of the process
- an analysis of the consequences of increased production for other parts of NIMWeC
- recommendations on the feasibility of selling into foreign markets

5.4 Identify sources of information

At an early stage, it is necessary to establish what access to information the analyst is going to have. This probably entails access to people working in the area, domain experts, documentation, observation, and so on. The political nature of the organisation also has an effect on how the analysts can go about the process of information gathering.

For NIMWeC, the client is an expert in all aspects of the operation of the company. The analysis is to be carried out by means of a series of interviews with this one person, together with a guided tour of the factory floor.

5.5 Define the contractual task

Much of the preliminary analysis is about reaching a shared understanding of the client's world, the developments envisaged and the analysis project to be carried out. A record of any agreement is usually necessary for a variety of commercial and legal reasons. This may include summaries of the points mentioned above, as well as estimates of cost and effort. Such a contractual agreement may be possible only after the process for the analysis project has been designed, an activity that is described in the next chapter.

The term *analysis project* has been used so far rather loosely to mean any kind of analysis endeavour. Once the stage of contractual agreement is reached, however, this agreement defines the analysis project as a particular piece of work to be carried out.

For NIMWeC, since the client seems like the right sort, we simply shake hands on the deal, roll up our sleeves, and set to work.

Chapter 6

Process Design

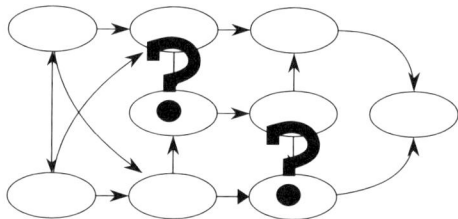

For NIMWeC, we choose to follow the Basic Process. Although this choice is convenient for expository purposes, the Basic Process is indeed suitable for the NIMWeC project. We have agreed in the preliminary analysis that we need to gain an understanding of the Old World, look at pathologies and prescriptions, and produce a specification of a New World NIMWeC that achieves its development objectives. The need to do all this suggests that we should be following the Basic Process, described in Chapter 4.

The Basic Process is summarised in Figure 4.1. We emphasise that the dependencies do not imply strict sequentiality. For example, 'model Old World' depends on 'gather information'. This does *not* mean that there has to be an Information Gathering 'phase' that has to take place before an Old World Purposive modelling 'phase'. Rather, the diagram should be read as 'Old World purposive modelling depends on gathering information'. In practice, modelling and gathering information happen side by side, with information gathering being driven (to some degree) by the needs of the modelling activities. Some information gathering may still be opportunistic, as sources become available to us or as observations are made.

If we were not to perform all of these activities, or if the situation were in some way different—if, for example, we were already well acquainted with the weaving process—then we would need to consider which parts of the ORCA process were needed. Now would be the time to do it.

There is no ideal or best process. A process that is effective on one project may be entirely inappropriate on another project. The nature of the process must reflect the nature of the problem being addressed, and the background of the people addressing the problem.

Although the ORCA process is split into three—preliminary activities, process design, process execution—the last two are not sequential: there is feedback from process execution to process design, in order to ensure that the process remains effective for the specific project. For example, if the domain is thought to be well understood, and a suitable tailoring is chosen, but unexpected complications are subsequently unearthed, it might be necessary to modify the process to include more Old World modelling. The designed process should include the means for handling such process redesign issues, as well as a plan covering products—both intermediate and deliverable—and the activities that generate and check those products.

The issue of process design is explored more fully in Part IV.

For NIMWeC, spurred on by an eager client, we move straight on to the task of understanding the Old World, with the plan of following the Basic Process until the job is done.

Chapter 7
Old World Purpose

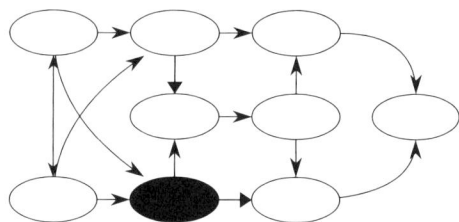

7.1 Introduction

In order to eliminate problems in the organisation, and to develop the organisation to achieve the objectives of the client, we need a thorough understanding of what the organisation is trying to achieve and how it currently goes about achieving it.

A good way of gaining such an understanding, particularly if the organisation is complex, is to model the Old World—the organisation as it currently is—using the ORCA modelling languages. The models that are produced should describe enough of the purposes and behaviour of the part of the world that makes up the organisation for us to understand what it is about, and should help to locate the cause of any current problems. In addition, attempting to achieve the development objectives is likely to have various consequences. The models will form the basis for understanding such consequences.

For this example, we do not expect to find major problems, since NIMWeC is currently working adequately, though not perfectly. The main aim of Old World modelling is to provide a context for future developments.

It should be emphasised that modelling the Old World is driven by our need to investigate and understand the situation. There is no notion that we are trying to produce a 'complete' model of the Old World. What is actually produced is at the discretion of the analyst, according to the circumstances and course of the analysis. All the modelling activities are iterative, with models being revised or extended in the light of subsequent investigations.

This chapter and the next describe the process of modelling NIMWeC's purpose and behaviour respectively. The analysis both of current problems and of the consequences of 'perturbing' the models with the development objectives is dealt with in Chapter 9.

In these chapters, we get to grips with the detail and terminology of the NIMWeC system. This may seem daunting at first, but it is necessary to show how ORCA

7.2 NIMWeC as a manufacturing organisation

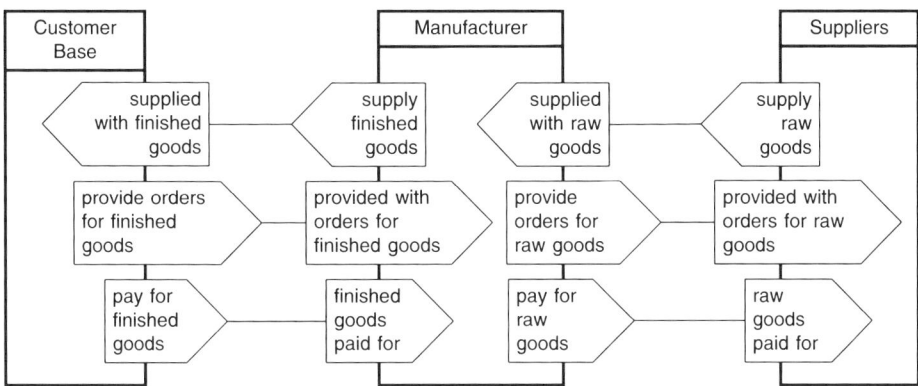

Figure 7.1 A simple manufacturing business

deals with a reasonably complex analysis situation. Reference should be made to Chapter 4 for background information on the example.

7.2 NIMWeC as a manufacturing organisation

Modelling the purposes of the organisation involves modelling the roles that make up that organisation. The roles are described in terms of how they co-operate to achieve purposes, and how their components allow them to achieve their purposes. The ORCA language for describing this is Grampus, illustrated in Chapter 2 and described more thoroughly in Appendix B.

Throughout this chapter, it should be remembered that we are talking about the service provision *arrangements* within the system. The behaviour that realises these arrangements is the subject of the next chapter.

7.2.1 From a general model to a model of NIMWeC

The initial modelling is done on the basis of the preliminary analysis (Chapter 5) and talks with the client. As is often the case in analysis, these are not the only sources of material. Here, recognising that NIMWeC is a manufacturing business, we bring to the project an idea of what a simple manufacturing business should look like: a business with both customers and suppliers (Figure 7.1). There are three *co-operations* between Manufacturer and its Customer Base: Manufacturer *guarantees* to supply finished goods to Customer Base; Manufacturer *relies* on Customer Base to provide requirements for the goods; Manufacturer relies on Customer Base paying for the supplied goods. In this ideal model, Customer Base's reliances and guarantees match Manufacturer's: Customer Base guarantees to provide the

Chapter 7 Old World Purpose

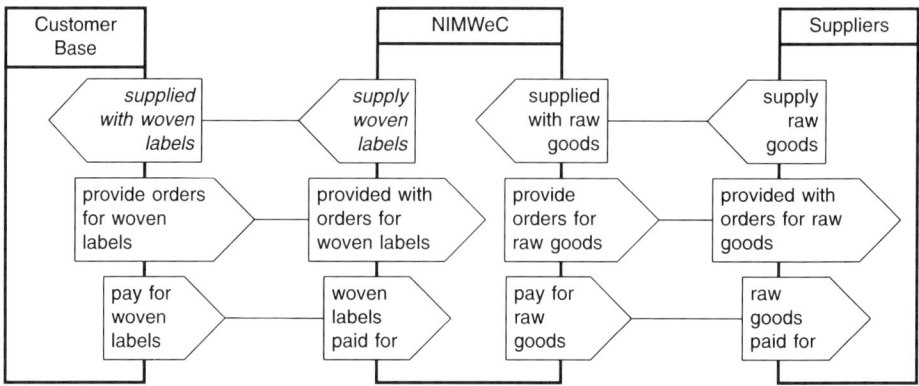

- Customer Base; *supplied with woven labels* in a timely manner
- NIMWeC; *supply woven labels* in a timely manner

Figure 7.2 A top level model of NIMWeC

requirements for finished goods, and to pay for them, and relies on Manufacturer to supply them. A similar collection of customer/supplier co-operations exists between Manufacturer and its Suppliers.

Within Manufacturer, we assume that some behaviour occurs which turns the raw goods that it relies on into the finished goods that it guarantees to provide. This is what makes it a manufacturing business.

We construct a top level model of NIMWeC based on this simple model, with the goods identified as being woven labels (Figure 7.2). The italicised guarantee and reliance names indicate that these have been *qualified*. The qualifier states requirements on the guaranteed or relied-upon service, here that the woven labels be supplied 'in a timely manner'.

When applying a general model to a specific situation, service descriptions need to be considered carefully to decide if they really are intrinsic to the roles to which they are ascribed. For example, if we were unsure about the suppliers' ability to supply the material that NIMWeC relies on, we would have to do further work—some form of market research—to establish what their intrinsic purposes are. Such work is outside the scope of this analysis, and so on the assurance of the client we show the services in this model as intrinsic.

Having identified the roles and co-operations at this level, the next step is to identify co-operations that seem to be of interest. Here, the most interesting co-operations, from what we can gather so far, seem to be those between Customer Base and NIMWeC (Figure 7.3). Some of the co-operations are *justified*. The justifications, labelled in the connector boxes and shown in the footnotes, are to do with the financial and contractual nature of the co-operations. This is indeed the extent of NIMWeC's dealings with its clients. The co-operations between NIMWeC

7.2 NIMWeC as a manufacturing organisation

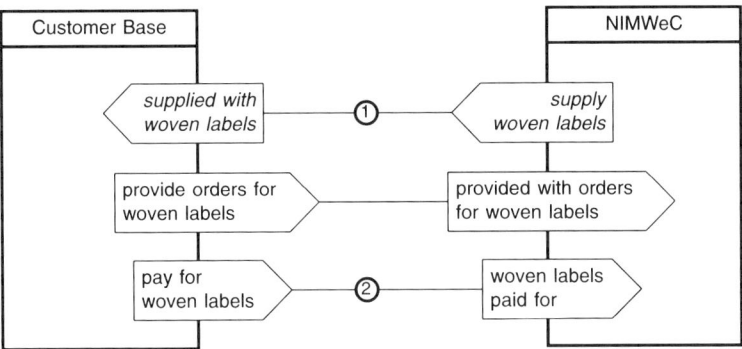

- Customer Base: *supplied with woven labels* in a timely manner
- Manufacturer: *supply woven labels* in a timely manner
- (1) labels must match the customers' requirements at reasonable cost to the customer, and at reasonable profit to NIMWeC
- (2) payment must fit into the cashflow requirements of both customers and NIMWeC

Figure 7.3 Co-operations between Customer Base and NIMWeC

and Suppliers do not seem so interesting; in fact, our client tells us that there are no problems or limitations with the suppliers, and so we pay no further attention to that role.

This done, the next stage of the modelling is to look more closely at each of the roles.

7.2.2 Investigating the surrounding roles

NIMWeC has two kinds of customer: name tape customers and shirt size label customers. Taken together, these constitute NIMWeC's customer base (Figure 7.4). In this case, the structure seems straightforward. The services of Customer Base are *delegated* to its subroles, Name Tape Customer Base and Shirt Size Label Customer Base. The co-operations between Customer Base and its two subroles are not in question, so the guarantees and reliances of the subroles are shown as being *intrinsic*.

Neither of the new subroles adds any new services. However, the nature of the co-operations with Customer Base is different in each case. The main differences are captured in the justifications of the co-operations. Since each higher level service is delegated to more than one lower level service, we have to demonstrate how the lower level services 'add up' to the higher level service. The justification points out the particular contribution that each delegated service makes to the fulfilment of the delegating service. So, for example, the reliance on Customer Base to pay for labels is fulfilled by both subroles. However, the nature of payment is different in

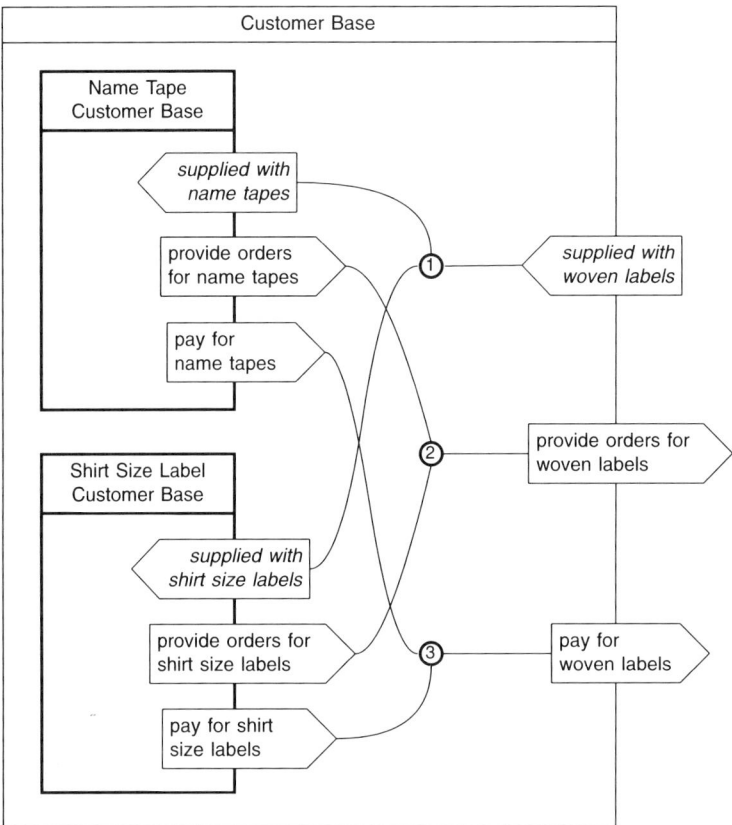

- Customer Base (and subroles): *supplied with woven labels* in a timely manner
- (1) The Name Tape customers require customised name tapes in a range of colours and and letter styles. The Shirt Size Label customers require a small range of shirt size labels.
- (2) Fulfilling name tape orders is profitable and the market has potential for growth. Fulfilling shirt size label orders is not profitable, but more than covers marginal costs and therefore contributes towards fixed costs. It allows production to be stable throughout the year, contributing to overall stability.
- (3) Payment for name tapes is in advance, with collection of payments being almost cost-free, and is highly seasonal. Payment for shirt size labels is requested by invoice, with an associated credit cost, and is constant throughout the year.

Figure 7.4 NIMWeC's Customer Base

each case, orders for name tapes being pre-paid and orders for shirt size labels being invoiced after they are woven. Together, these two payment methods satisfy the need for labels to be paid for. The combination has the additional complementary property that the payments balance out over the year. Payments for name tapes arrive during the summer whereas those for shirt size labels arrive more regularly throughout the year, but principally during the winter which is when most of the shirt size label weaving is done.

Here, the justification is given in quite plain terms. Later, we see a more structured argument.

7.3 Investigating NIMWeC

So far, we are happy with the simple high level view that we have of the roles that NIMWeC interacts with. The next step is to look inside NIMWeC itself. To do this, we need to decide on the following:

- what roles make up NIMWeC
- what the intrinsic guarantees and reliances of these roles are
- how the guarantees and reliances of the higher role are delegated to the component roles, and the justification of the delegations
- how the component roles co-operate, and the justification of the co-operations

Few of these decisions can be made in isolation; making one decision can limit the possibilities for other decisions, or demand the making of new decisions. Moreover, the decision-making process is an iterative one; making one decision may affect the validity of previous decisions. This aspect of the process is not discussed further.

7.3.1 Choosing a structure

Choosing roles that constitute one level of abstraction has so far been guided either by our prior knowledge of manufacturing (giving us Customer Base, Manufacturer and Supplier) or by explicit statements given to us during the preliminary analysis (giving us Name Tape Customer Base and Shirt Size Label Customer Base). Now, we face some more difficult decisions about the way in which to model NIMWeC's *formation* (a role's structure in terms of subroles, promotions and delegations).

There may be many possible formations of any given role. What we are looking for in choosing between different formations is their explanatory power. For example, it would be possible to form NIMWeC from two roles: a shirt size label production role and a name tape production role. However, this would not be satisfactory for our purposes. NIMWeC is focused on the production of name tapes, and the production of shirt size labels is a means of supporting this. Separating these two activities would not help in understanding NIMWeC. Another kind of formation could be based on its financial components, perhaps identifying those units that raise revenue, those that dispose of it and those that control the flow of revenue. However, the client has excluded the issues related to finance from our terms of reference.

The formation we choose is a basic one, reflecting the two major *activity areas* of the company—order processing/invoicing, and manufacturing. It is important to note that the way we choose to model the company in terms of roles may or may not coincide with the actual organisational structure. The roles and co-operations are a rationalisation of NIMWeC. In practice, an organisational division may fulfil many roles; a single role may be spread across several divisions. As it happens, the organisation of NIMWeC is basically by activity area, along the lines that we are suggesting.

7.3.2 Constructing intrinsic services

Deciding on intrinsic guarantees and reliances involves a consideration of what the role is *intended* to do without regard, for the moment, for any other role. For NIMWeC, the services are derived from the preliminary analysis and from further talks with the client.

Manufacture in NIMWeC is responsible for the production of name tapes, and relies on the provision of weaving requests. Orders & Invoicing is responsible for taking customer orders for woven labels, providing weaving requests and collecting payments. It relies on a customer base that both places orders and makes payments, and also relies on the provision of woven labels (Figure 7.5).

7.3.3 Constructing delegations and promotions

The process of constructing delegations and promotions involves the following two steps:

- identifying the component roles as targets for the higher level guarantees and reliances
- matching each higher level guarantee and reliance with the guarantees and reliances of the component roles, and justifying the resulting delegations

Identifying the target roles is straightforward for NIMWeC. It has one guarantee and two reliances. Each of these can be associated with the Orders & Invoicing role, and is shown (temporarily) by placing an *extrinsic* service description on it (Figure 7.6). This identification is straightforward, since the formation is based on the activity areas of NIMWeC, and these activity areas are reflected in the services we ascribe to NIMWeC. That is not to say, however, that it is an automatic process.

Next the extrinsic guarantees and reliances have to be matched with intrinsic services. Again, we find that this is a straightforward step for NIMWeC. In particular, we decide that the two NIMWeC reliances have direct counterparts

7.3 Investigating NIMWeC

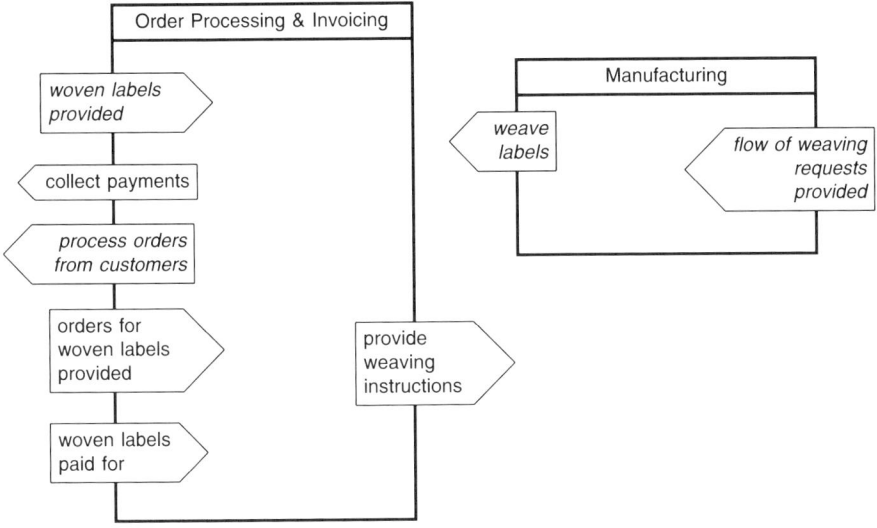

- Manufacturing:
  - *weave labels* accurately, to requirements, in a timely and cost-effective manner
  - *flow of weaving requests provided*: the requests are well ordered (minimising down-time due to colour changes), and frequent (minimising down-time due to lack of instructions, but not so frequent that name tapes cannot be woven in a timely manner)
- Orders & Invoicing:
  - *woven labels provided* in a timely manner, and in a way that they can be matched with their order
  - *process orders from customers*: in a timely manner

Figure 7.5 Manufacturing and Orders & Invoicing

in Orders & Invoicing, and our confidence in these delegations is reflected in the lack of a justification. The delegation of the NIMWeC guarantee is more complex, however, and a justification is given which emphasises the role that each service plays in the fulfilment of NIMWeC's main business (Figure 7.7).

7.3.4 Constructing co-operations

The process of constructing co-operations is similar to that for delegations and promotions, and involves the following similar steps:
- for each role, identifying roles as service providers or receivers
- matching each extrinsic service description with one or more intrinsic service descriptions of the identified roles, and justifying the resulting co-operation

62 Chapter 7 Old World Purpose

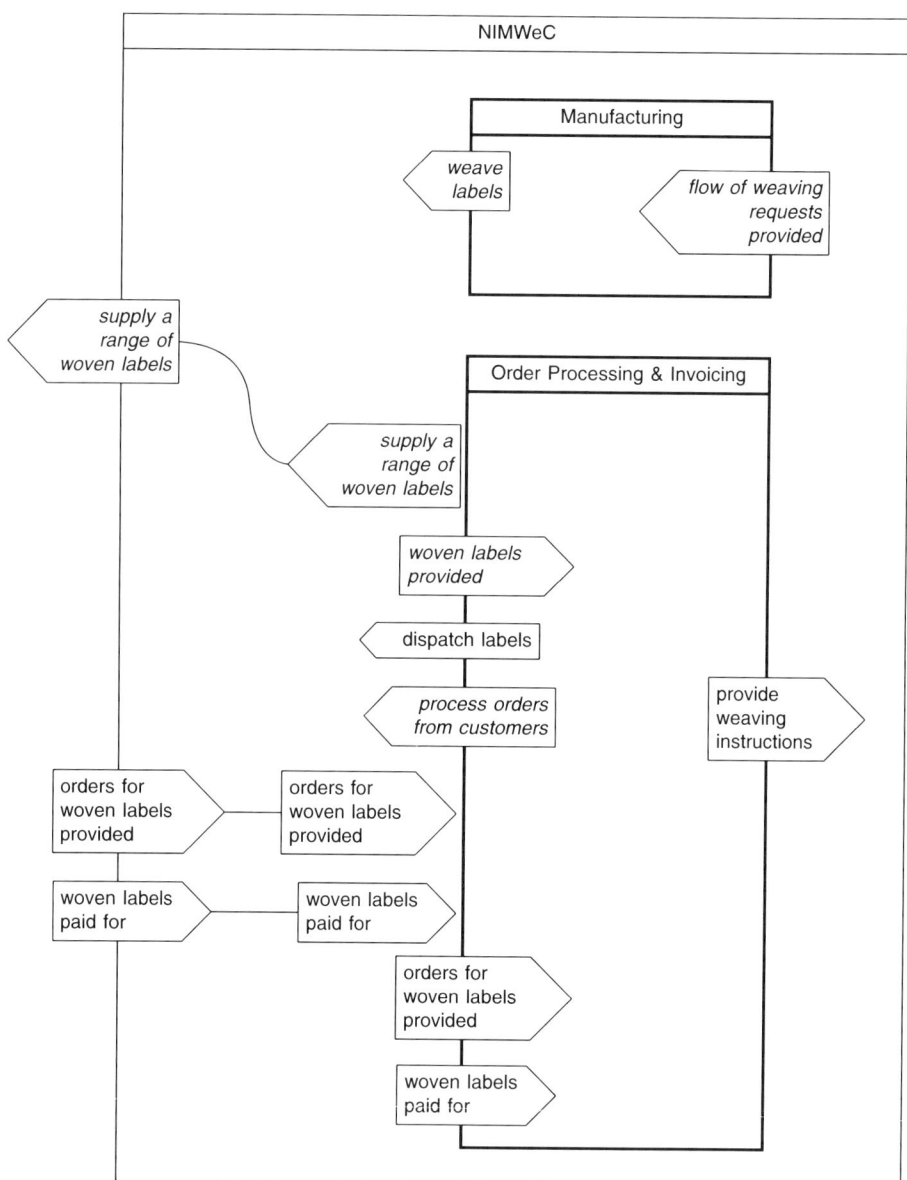

- qualifiers as in Figure 7.5

Figure 7.6 NIMWeC's delegations and promotions

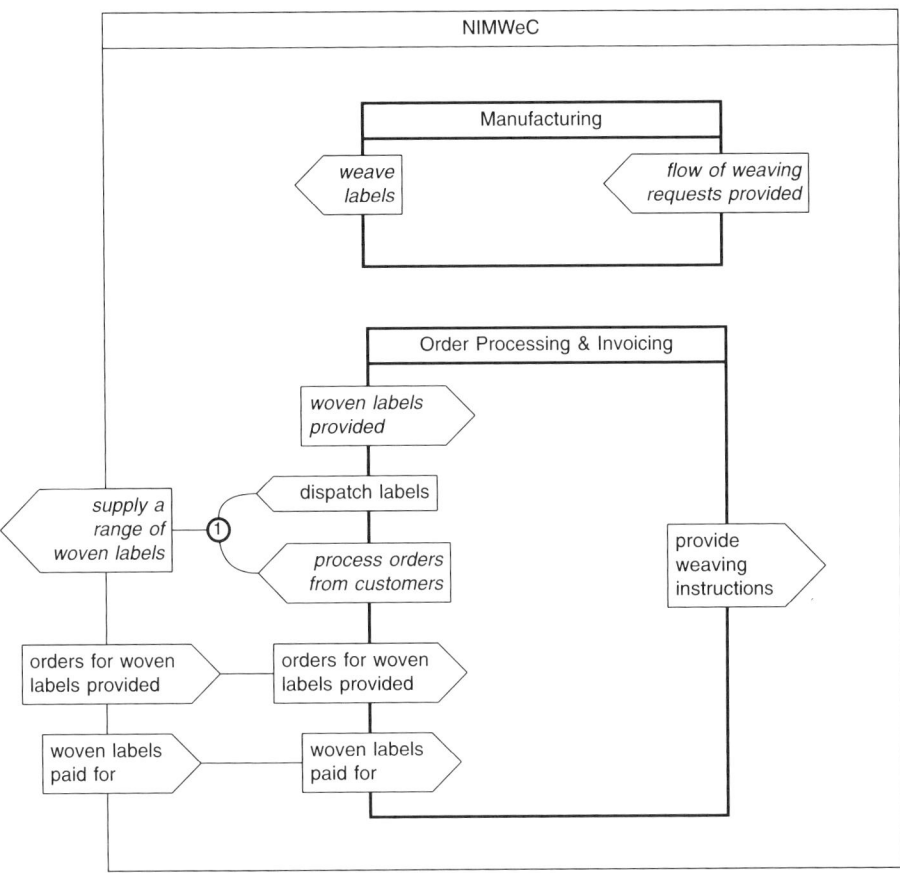

- qualifiers as in Figure 7.5
- (1) NIMWeC manufactures labels for sale in response to customer requests

Figure 7.7 Matching NIMWeC's delegations and promotions

For Orders & Invoicing, the only possible target role in the current model is Manufacture. Its reliance on the weaving of labels seems to be obviously met by the Manufacture guarantee to weave labels. Each service, however, mentions the idea of 'timely' production of name tapes. It is important that the same definition of 'timely' applies to each service. We decide that as far as we know the same definition is used, but make sure to mention this assumption in the justification of the co-operation (Figure 7.8).

If NIMWeC were interested only in supplying the name tapes and not in weaving for its own sake, the possibility of solutions such as out-sourcing the weaving to other weaving companies could be considered. The fact that it is not is captured by its co-operation with Manufacturing.

64   Chapter 7   Old World Purpose

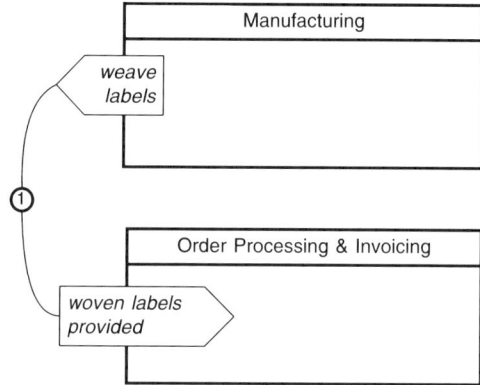

- qualifiers as in Figure 7.5
- (1) weaving must be done in such a time that Orders & Invoicing can meet its timeliness qualifier

Figure 7.8 The 'weave labels' co-operation

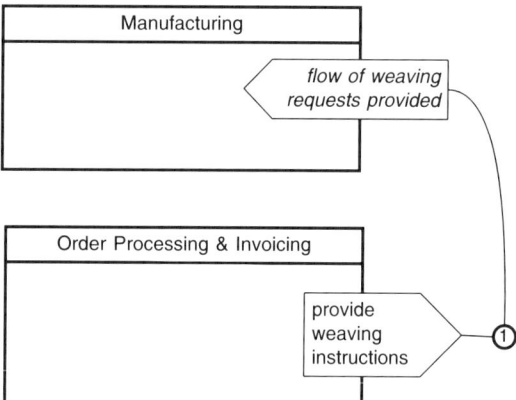

- qualifiers as in Figure 7.5
- (1) Orders & Invoicing is responsible for ensuring that there is a proper flow of weaving instructions

Figure 7.9 The 'provide weaving instructions' co-operation

7.3 Investigating NIMWeC

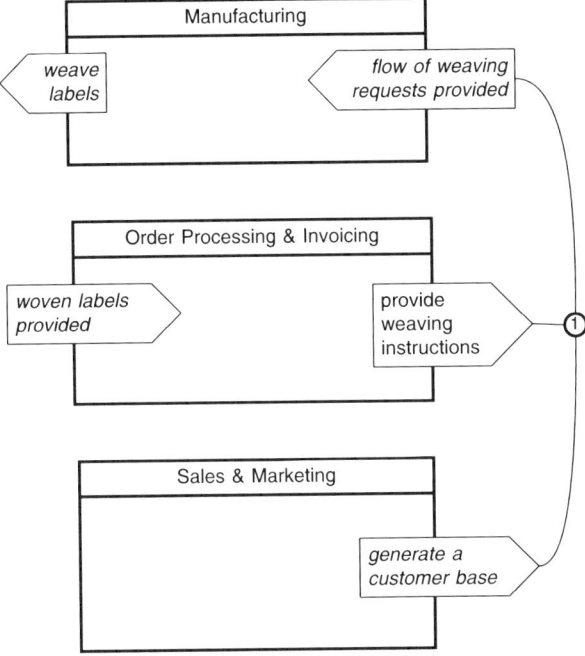

- Manufacturing:
  - *weave labels* accurately, to requirements, in a timely and cost-effective manner
  - *flow of weaving requests provided*: the requests are well ordered (minimising downtime due to colour changes), and frequent (minimising idle-time due to lack of instructions, but not so frequent that name tapes cannot be woven in a timely manner)
- Orders & Invoicing: *woven labels provided* in a timely manner, and in a way that they can be matched with their order
- Sales & Marketing: *generate a customer base* so that the customer base will provide orders in the future
- (1) Orders & Invoicing is responsible for ensuring that there is a proper flow of weaving instructions in the short term. Sales and & Marketing is responsible for generating a customer base, ensuring that there are orders in the long term.

Figure 7.10 The amended 'provide weaving instructions' co-operation

Orders & Invoicing's guarantee to provide weaving instructions seems at first sight to match the Manufacture reliance on a flow of weaving instructions (Figure 7.9). Thinking hard about the nature of a flow of weaving requests, however, we understand that Orders & Invoicing provides only a short-term flow of requests, and that we have missed out a crucial aspect of the NIMWeC business—Sales & Marketing. This is the role that generates the customer base in order to ensure that NIMWeC can rely on orders. Our initial thought is that this generation of the customer base will be a guarantee made to Orders & Invoicing. The Orders & Invoicing reliance on the provision of orders would be met both by this guarantee from Sales & Marketing and, through delegation, by the guarantee to NIMWeC from Customer Base to provide orders for woven labels. However, when we try to confirm this with the client, we are told that really the heart of the business is Manufacture, and the need for a long-term flow of orders lies in the need to keep on weaving, not to keep on taking orders. We decide to model the co-operation accordingly, and amend the model (Figure 7.10). The co-operation's justification demonstrates the complementarity of the two guarantees, which together ensure a flow of weaving requests both in the short term and in the long term.

Having introduced this new role, we need to see how it fits in with the rest of our model of NIMWeC. We run through the process of deciding on its intrinsic guarantees and reliances, deciding on the delegations and co-operations that it is involved in. The only other service we come up with for Sales & Marketing is a reliance on the customer base that is amenable to manipulation. Since we are making changes in the middle of a role, we need to consider the construction of delegations not simply as a top down activity, but a 'middle out' one. The new service we have added may in turn delegate to a subrole of Sales & Marketing, but it also propagates up to NIMWeC, and becomes a reliance on Customer Base.

Our final model of NIMWeC at this level is shown in Figure 7.11. To complete the tracing through of the propagation we amend the higher level model, placing a guarantee on Customer Base to match NIMWeC's new reliance. This in turn is delegated to its two component roles, Name Tape Customer Base and Shirt Size Label Customer Base. Since this higher level model was derived from our generic manufacturing model, we also think about amending our generic manufacturing model to include this new co-operation. However, rather than changing our generic model straight away, we decide to wait until we see how useful the co-operation is in aiding our understanding of this particular manufacturing organisation.

7.4 Further investigation

So far we have a model of NIMWeC with three component roles. As with our original high level model, we can decide whether to model any of these in greater detail, and, for each, go through the formation process as outlined above. We are sure that we don't want to look any further at the moment at Sales & Marketing, important though this is to NIMWeC. We do feel that we need to look more closely at the other roles, and it is likely that any problems resulting from consequences of the development objectives will affect roles at this lower level.

7.4 Further investigation   67

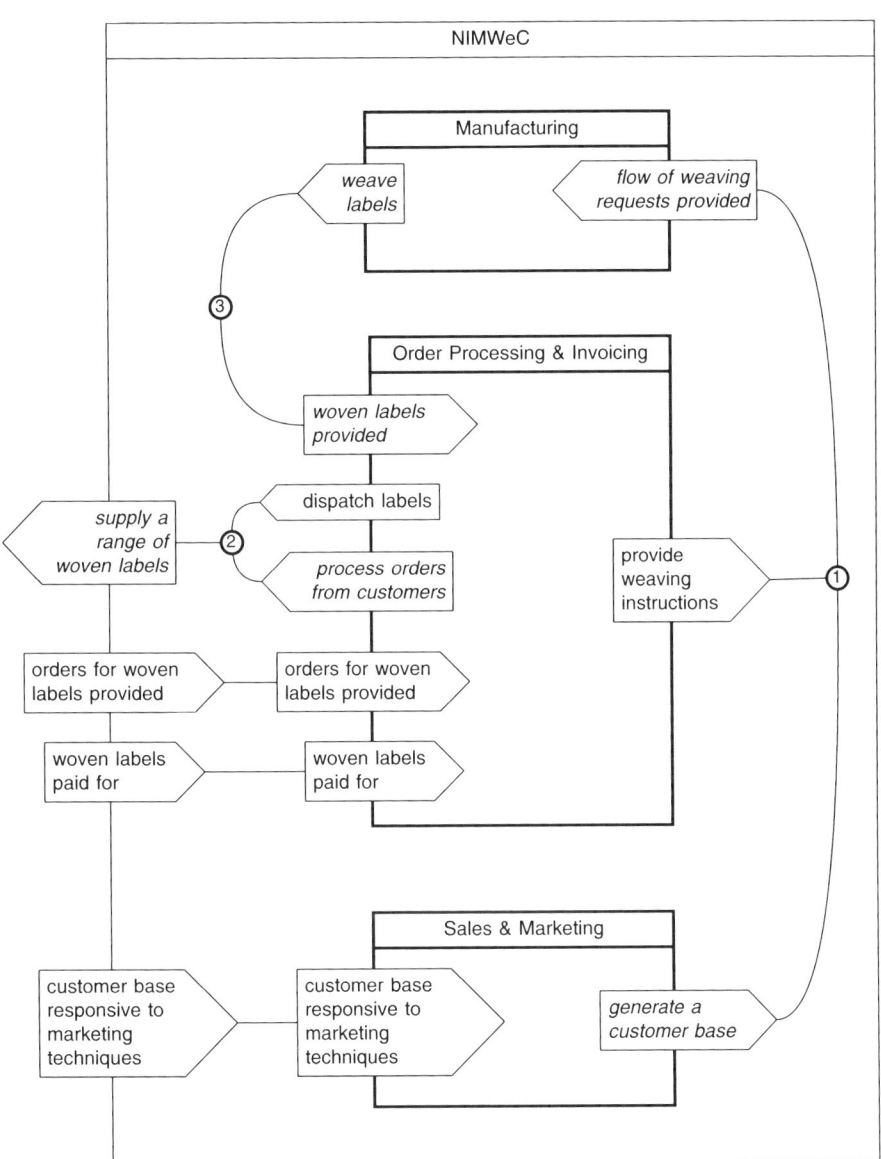

- Qualifiers and justifications as given in preceding figures

Figure 7.11 Detailed NIMWeC model

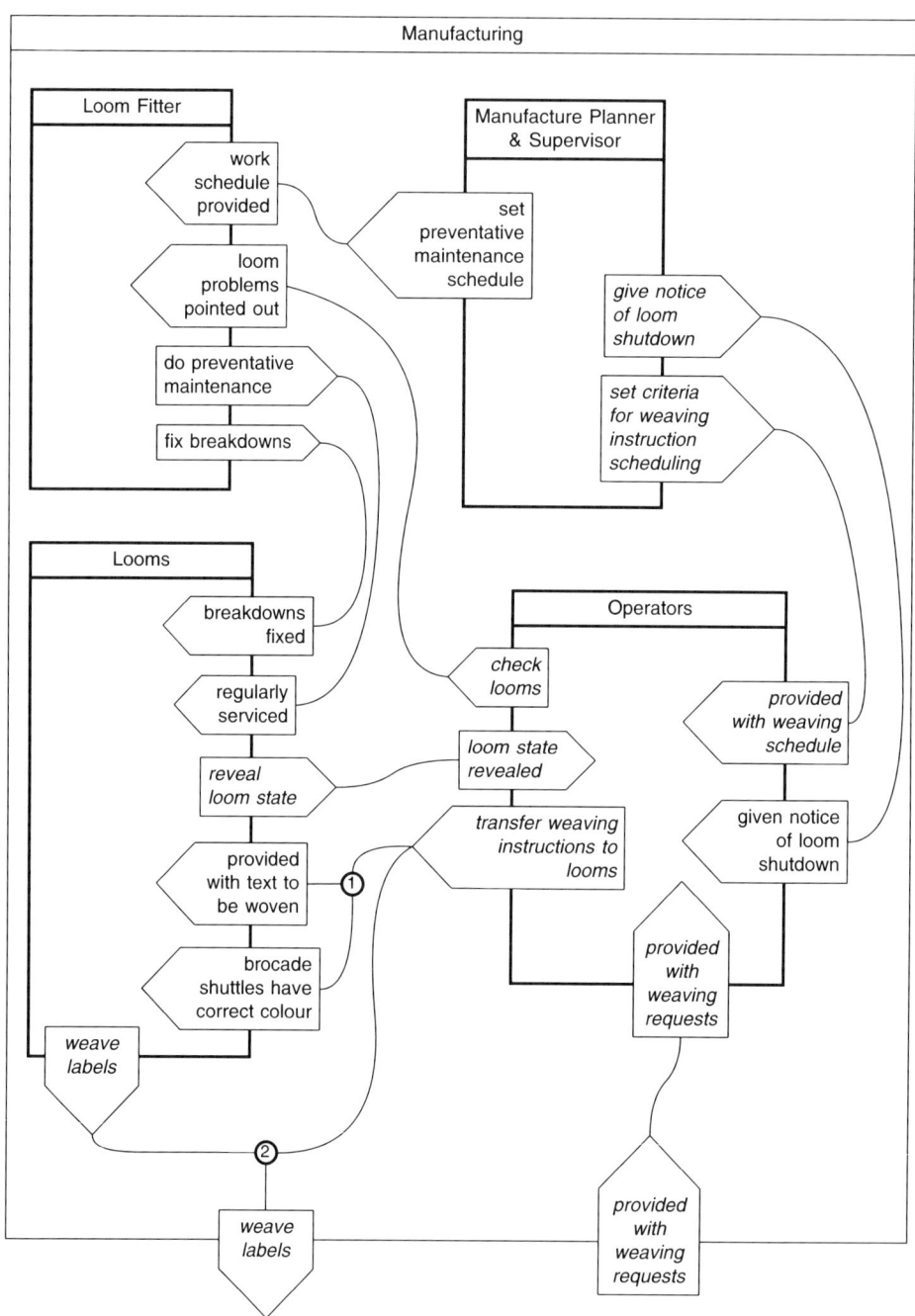

- Qualifiers and justifications are given in Figure 7.13

Figure 7.12 Detailed Manufacturing model

- Looms:
    - *reveal loom state* as required by the operators
    - *weave labels* according to instructions, with low down-time due to breakdown or difficulty of set-up, with efficient use of resources
- Operators:
    - *check looms* for loom breakdown, weaving accuracy, resource requirements
    - *loom state revealed* to allow detection of loom breakdown, weaving accuracy and resource requirements
    - *transfer weaving instructions to looms*: the instructions are well ordered (minimising down-time due to colour changes), and frequent (minimising idle-time due to lack of instructions, but not so frequent that name tapes cannot be woven in a timely manner)
    - *provided with weaving schedule* about change-over from name tape to shirt size label weaving
    - *provided with weaving requests*: the requests are well ordered (minimising down-time due to colour changes), and frequent (minimising idle-time due to lack of instructions, but not so frequent that name tapes cannot be woven in a timely manner)
- Manufacture Planner & Supervisor:
    - *give notice of loom shutdown* for preventative maintenance, and for change of warp
    - *set criteria for weaving instruction scheduling* concerning minimum operational time between colour changes, and allocation of looms to name tape or shirt size label production
- (1) transferred weaving instructions include both colour set-up and required text
- (2) acceptable label weaving requires both loom capability (for accuracy) and loom operation and maintenance (for timeliness)

Figure 7.13 Qualifiers and justifications for detailed Manufacturing model

7.4.1 Manufacturing

The model we produce for Manufacturing is shown in Figures 7.12 and 7.13. The higher level reliance on a flow of weaving instructions is promoted from the Operator's reliance, and the higher level guarantee to weave is delegated to both the Machine and the Operator. The justification of the latter emphasises the need for the machine to function both correctly and reliably in order to weave labels in the desired manner.

Broadly, the co-operations are to do either with the maintenance of the Loom role, or the operation of the Loom role.

Loom maintenance:

- Loom Fitter guarantees to fix the Loom when it breaks down, and the Loom relies on breakdowns being fixed. In order to carry out this guarantee, Loom Fitter relies on problems with the Loom being pointed out. In order to check the looms, and so point out problems, Operator relies on the information necessary to be able to check the loom being provided. Loom guarantees to

provide this information.
- Loom Fitter guarantees to provide maintenance, and Loom relies on being maintained; Manufacture Planner & Supervisor guarantees to provide a maintenance schedule to Loom Fitter.

Machine operation:
- Manufacture Planner & Supervisor provides shut down instructions and criteria for the large-scale scheduling of weaving instructions.
- Loom relies on being provided with weaving instructions for text and style, and set-up instructions for colour; Operator guarantees to transfer the weaving instructions to Machine, minimising Machine idle- and down-time, in a reasonable time, and in such a way that the woven name tapes can be matched with their orders.

It is the last of these co-operations that is most interesting, not because it is in itself problematic, but because it locates some of the behavioural problems in NIMWeC. We are now at the level of the interface between Operator and Loom, and there is not a great deal more to be said about Manufacture at the purposive level—and much of what we have said already can be construed as abstract descriptions of behaviour. It is time to stop purposive modelling in this area.

7.4.2 Orders & Invoicing

Orders & Invoicing can be formed from two roles, corresponding to the Name Tape production and Shirt Size Label production. This formation is similar in style to that for Customer Base, and the comments made regarding the complementarity of those two roles with regard to the delegated services also apply here. We decide that we are not interested in the ordering process for shirt size labels, since the client tells us that:
- depending on how the development of the name tape side of the business goes, this aspect of the business may disappear entirely
- the ordering process for shirt size labels is small scale and flexible, and it would take a great deal of change to harm or improve it significantly

We could choose a formation of Name Tape Orders & Invoicing based again on the major activities of ordering, dispatch and payments. But these look more like aspects of behaviour than purposive roles. So, as with Manufacturing, the behaviour of these services has become more interesting than their co-operations, and it is time to stop purposive modelling.

Chapter 8

Old World Behaviour

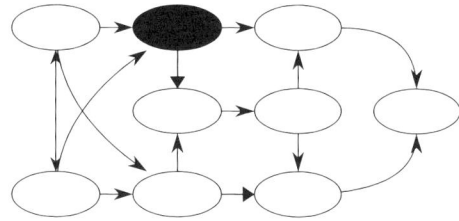

8.1 Introduction

Before we can understand the consequences of the development objectives, we need to understand how the services identified in the previous chapter are supported by actual behaviour. If the current behaviour is adequate to meet these needs it may be that the development objectives have no major consequences at all. It is more likely, however, that some change in the current behaviour is needed. In either case, we need a good understanding of the current behaviour in order to make such assessments (which we do in the next chapter).

For NIMWeC, there are two particular areas that we need to know more about— the operation of the weaving machinery and the processing of orders for name tapes. These seem likely to be affected by the development objectives. Also, we don't yet have a good understanding of what currently happens.

Within the area of machine operation we are interested in the machine's interface with the operator, the product (what the name tapes look like) and the weaving process itself which lies at the heart of NIMWeC. Within the order processing area we are interested in orders, batches and the process that constructs batches from orders.

The link between these two areas is identified in the purposive models as a co-operation, with Manufacture relying on the provision of suitable weaving instructions, and Orders & Invoicing guaranteeing to provide these. We now need to find out what a weaving instruction is, and the nature of the co-operation. In other words, the *service* that the co-operation concerns (described informally in the purposive model) needs to be modelled more formally as an aspect of behaviour, using the Beluga language.

Although we need to understand the current behaviour of machine operation and order processing, we need to be careful not to model anything in too much detail at this stage, since things may well be different in the New World.

As we suggested at the end of the last chapter, it is not sensible to model these areas using Grampus, since it is their behaviour that is of interest—there is no more to be said about their purposes. In this chapter we use ORCA's behavioural modelling language, Beluga, together with other techniques as appropriate. The Beluga modelling language is explained in detail in Chapter 12, and formal definition of the language constructs is given in Appendix C.

The starting point for Old World behavioural modelling is simply the areas as identified in the purposive models—there is no need to continue the decomposition of purposive entities. This approach was useful in dealing with the make-up of organisations, but is not necessarily useful when modelling behaviour in terms of objects and interactions.

The end point for Old World behavioural modelling occurs when we feel we have enough of an understanding to support the identification of pathology and prescriptions (see Chapter 9).

In reading the remainder of this chapter, reference may be made to Section 4.3 for a summary of the gathered information and a glossary of weaving terms.

8.2 The weaving of name tapes

The first thing we notice when we walk into the factory is the noise. Then we notice the looms. There are two of these, each with one operator. Strips of name tapes come out of one side of a loom, while the operator stands at another side and alternately pummels a board with round pegs and consults a sheet of card. Every now and then the operator walks around and cuts off the strips, and puts them into a bag along with one of the sheets. The operator then takes the bag to one side of the factory floor and stacks it with some others. Next to this stack is a pile of the sheets. The operator, while over there, shuffles through the sheets, selects one, and heads back to the loom.

After further discreet observation, we question the operators and the client about what exactly is going on...

8.2.1 The weaving process

The thing that strikes us most about the weaving process is how everything seems to be geared around the capabilities of (or limitations of) the loom itself and its associated control mechanisms, and so we feel it might be worth studying the behaviour of these.

A model of static structure

We start by trying to identify the structure of the looms, which we are told are nearly identical. In what follows, it is important to remember that we are investigating the behaviour that realises the services of an already identified role, rather than partitioning the role further into a number of subroles as we were doing for the most part in our purposive modelling. We are opening a box of chocolates to discover its contents, rather than cutting a birthday cake in order to share it around. The information we have to go on is as follows.

The major part of the loom is really three separate parts: a weaving frame, two jacquards and a lifting box:

- The *weaving frame* is physically connected to two jacquards (which can be regarded as one 'jacquard thing', since each does identical things here) and a lifting box.
- The *jacquard*, by means of hooks, lifts and drops warp threads so as to form the correct shed (the pattern of raised or unraised warp threads) for the shuttles carrying the weft threads to pass through.
- The *lifting box* lifts into place either a ground shuttle or a brocade shuttle for each ribbon, depending on what is to be woven: the body of the ribbon (a ground pick) or a pattern on the ribbon (a brocade pick).

The other parts of the loom, we are told, are concerned with the control of this machinery:

- The *cam shaft* is physically connected to the weaving frame, and rotates once through 360 degrees for every machine cycle, or *pick*, in which the correct shed is formed and the shuttles passed through the warp threads.
- The *cam contacts* are attached to the control box, and detect the degree of revolution of the cam shaft.
- The set of *electromagnets* (one for each warp thread) is attached to the jacquard, and, on receipt of complex *turnOn* or *turnOff* messages from the control box, causes some of the *jacquard hooks* to be lifted or dropped.
- The *lifting box solenoid* is attached to the lifting box, and, on receipt of simple *turnOn* or *turnOff* messages from the control box, causes the lifting box to either lift or drop the shuttles for each ribbon.
- The *control box* has states that are dependent on the information from the cam contacts. It sends messages to the electromagnet and lifting box solenoid at appropriate times during the pick, and allows an operator to set up its *patch panels*.

We can make this arrangement clearer by modelling the different classes of *Loom* component, and the interactions that connect the various components (Figure 8.1). Each box represents a *class* of Loom component. In some cases (Weaving Frame, Cam Shaft, Control Box), there is only a single instance of a class; in other cases (ElectroMagnet, Lifting Box Solenoid, Cam Contact), there are many instances. Some boxes contain the names of key operations for a class. An arrow between two boxes indicates that instances of one class interact with (operate on) instances of the other class. For example, Jacquards operate on the Weaving Frame.

74   Chapter 8   Old World Behaviour

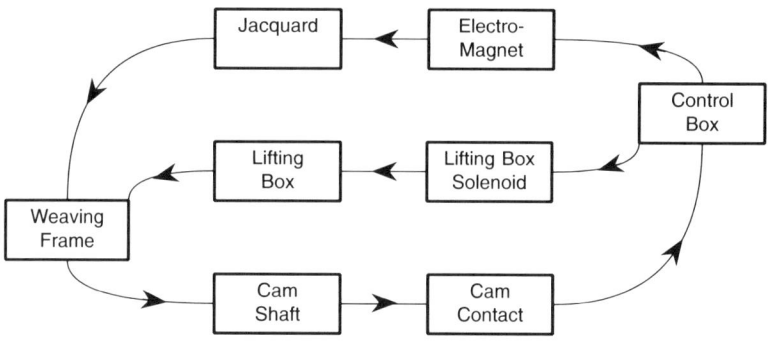

Figure 8.1 Loom model

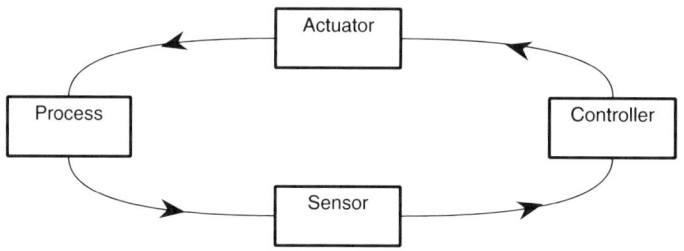

Figure 8.2 A Control System

This rather flat picture does not yet capture the distinction that we were told about, between the machinery and the control mechanisms. However, we have some knowledge of what a typical *control system* looks like (Figure 8.2). In a *ControlSystem*, the *Actuators* act on the *Process* according to instructions from the *Controller*, and the *Sensors* inform the *Controller* of the state of the *Process*. The *Process* is assumed to produce some product; in the case of a loom, it is name tape ribbons.

Armed with this knowledge, we can look for the elements of a control system in our model of *Loom*. In fact, at this level we can find three possible ways of describing the weaving process as a control system, depending on how we group the classes identified above.

The first grouping treats the weaving frame as the controlled process; the control box, electromagnets, lifting box solenoids, and cam contacts comprise the controller; the jacquards and the lifting boxes comprise the actuators; the cam

8.2 The weaving of name tapes   75

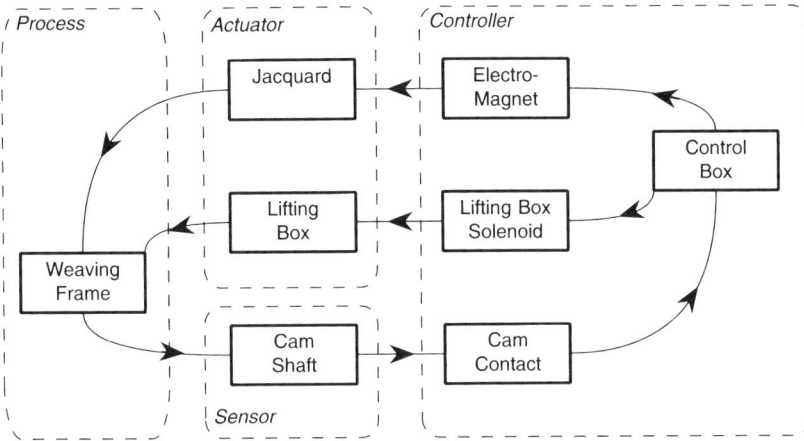

Figure 8.3 A mechanical partitioning of *Loom*

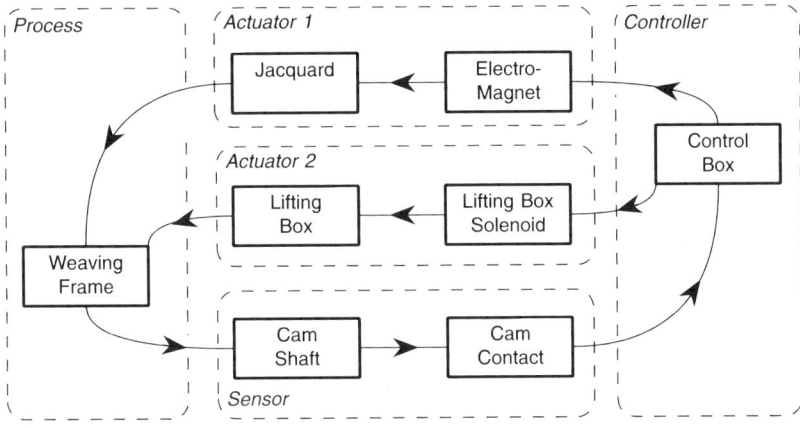

Figure 8.4 An alternative partitioning of *Loom*

shaft comprises the sensor (Figure 8.3). This approach focuses on the mechanical side of the weaving process, treating anything electromechanical as being hidden within the controller (and so of no interest).

The second grouping treats the control box as the controller, and the weaving frame as the controlled process, but bundles the jacquards with the electromagnets, and the lifting boxes with the lifting box solenoids as the actuators, and bundles the cam contacts with the cam shaft as the sensors (Figure 8.4). This approach focuses on the weaving frame and control box, treating the actuators and sensors as complex mechanical and electromechanical devices.

The third grouping, which seems to be the most appropriate given our interest

76   Chapter 8   Old World Behaviour

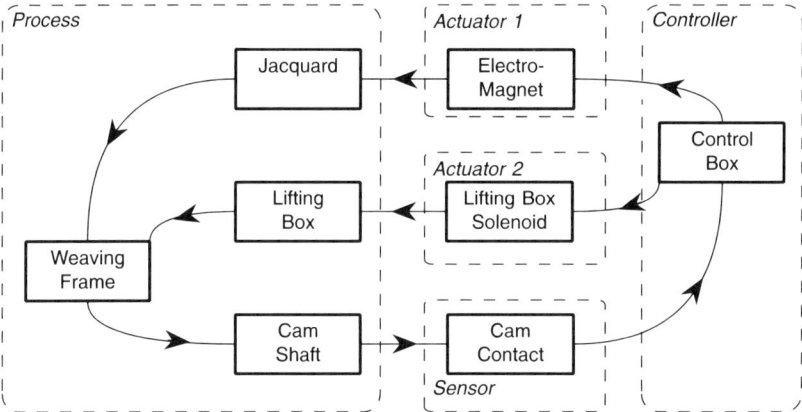

Figure 8.5 A physical partitioning of *Loom*

in the control box, is as follows: the control box is the controller; the electromagnets and the lifting box solenoids are the actuators; the cam contacts form the sensor; the jacquards, lifting boxes, cam shaft and weaving frame together form the controlled process (Figure 8.5). This approach bundles up the physical aspects of the loom, leaving its interface to the controller at the level of the electromechanical devices. This fits in with our interests—we are not interested so much in the physical aspects of the loom as in the demands it places on the control box through its electromechanical interface.

A model of dynamic behaviour

So far we have some idea of how these bits all work together, but in order to model the dynamics we need more information. The client gives us some old documentation for the loom, which contains diagrams showing the timing of one complete machine cycle (one pick), with respect to the electromechanical devices: the cam contacts, the electromagnets and the lifting box solenoid. The timing diagram for the current pick is shown in Figure 8.6 for the case where the following pick is to be a brocade pick.

We can see from this timing diagram that the machine progresses through six states within this one pick, as measured by one revolution of the cam shaft:

state 0 cam contacts 1, 2 and 3 are open

state 1 contact 1 is closed; the electromagnet is on (that is, some subset of the individual electromagnets)

state 2 contact 1 is open again; the electromagnet is off

state 3 contact 2 is closed; the lifting box solenoid is on

state 4 contact 2 is open again; the lifting box solenoid stays on

8.2 The weaving of name tapes

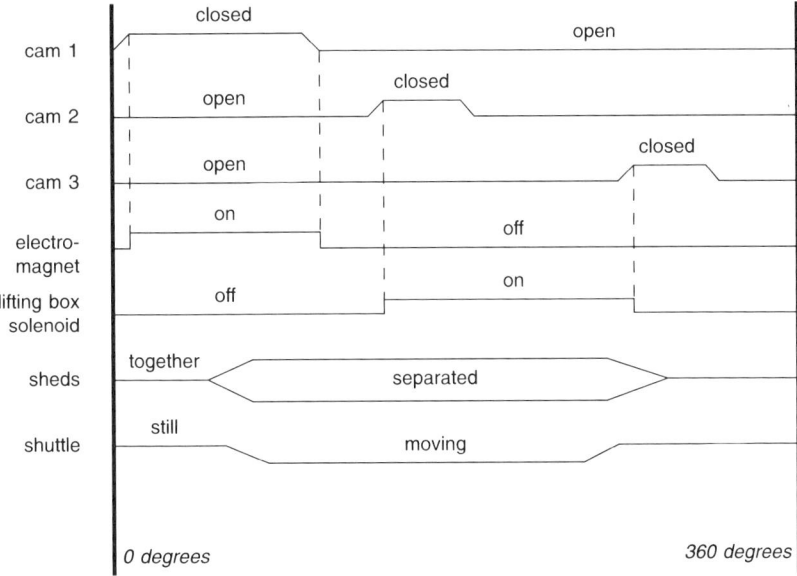

Figure 8.6 Timing diagram for the current pick

state 5 contact 3 is closed; the lifting box solenoid is off

If, instead, the following pick is to be a ground pick, then in the current pick the lifting box solenoid remains off and the lifting box does not move. The reason for the delay, we are told, is that the lifting box controls the mechanism that automatically moves the ribbons forward. If the lifting box is not raised, then the ribbon is moved on automatically for the next ground pick to be woven. If it is raised, then the ribbon is not moved on: the next pick is to be a brocade pick, with the brocade weft being inserted on top of the previous pick's ground weft. This prevents gaps in the name tape ribbon. There are thus two versions of states (3) and (4), depending on whether the Lifting Box Solenoid is on—(3a) and (4a)—or not—(3b) and (4b).

With the information from the timing diagrams we can build a dynamic model of a loom. The behaviour of *Loom* is a succession of *OnePick* behaviours (that is, complete machine cycles). Within *OnePick*, we can show how the various components interact (Figure 8.7). This Beluga diagram is similar to the timing diagram provided to us by the client (the Beluga *timelines* progress down the page, rather than left-to-right). In this dynamic model we are dealing with *instances* of the classes identified earlier:

- three *CamContacts*, named *cam*1, cam2 and *cam*3
- one *ControlBox*, named *cb*

78   Chapter 8   Old World Behaviour

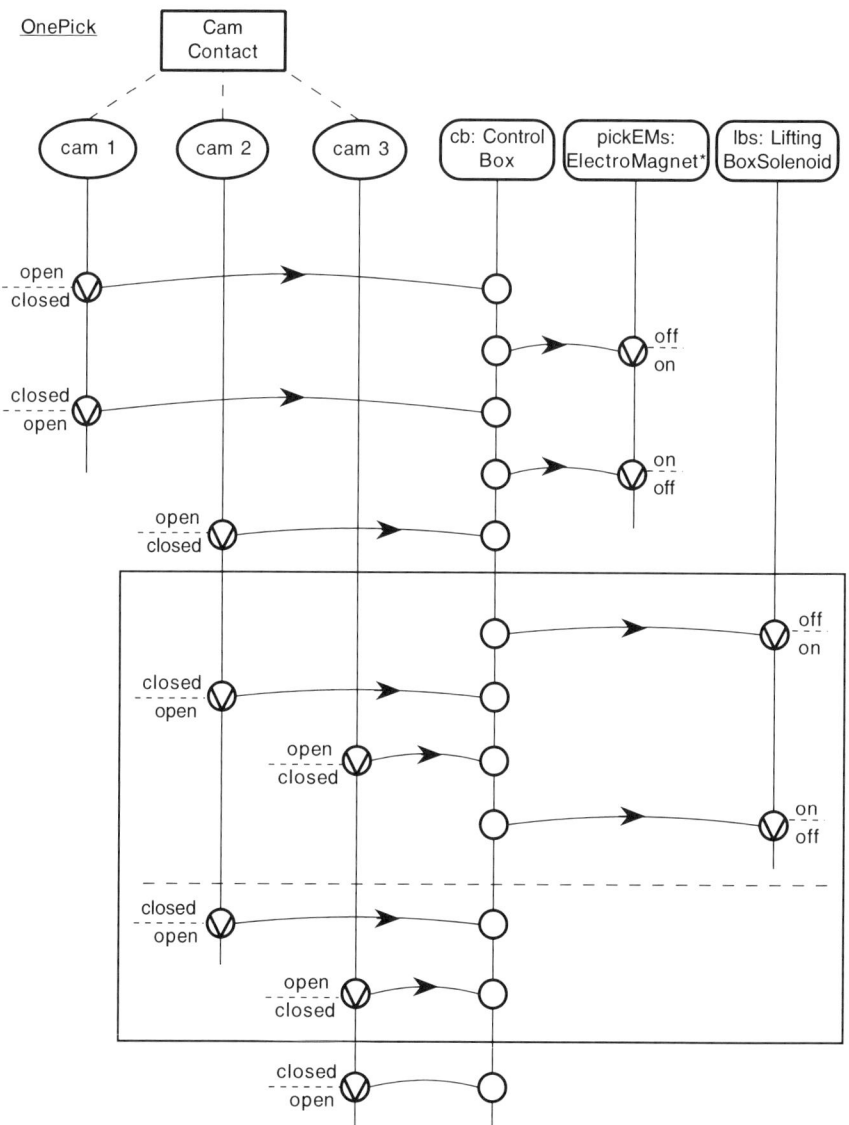

Figure 8.7 A dynamic model of *Loom*—the *OnePick* behaviour

8.2 The weaving of name tapes   79

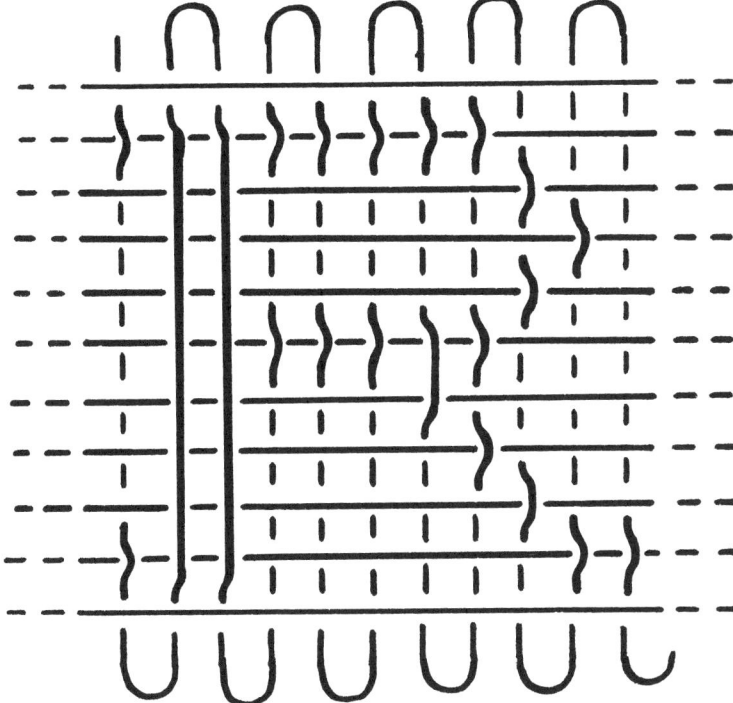

Figure 8.8 Brocade pattern for the letter 'R'

- one *LiftingBoxSolenoid*, named *lbs*
- a subset of the *Electromagnets*, named *pickEMs*

Each node on the timeline for a particular set of constituent objects denotes involvement of those objects in some bit of behaviour. Arrows linking nodes on different timelines show interaction *events* (that is, instances of the interaction relationships between classes shown in Figure 8.1). Changes in the state of either source or target objects is shown with the 'before' state above the 'after' state. Two alternative episodes in a machine cycle are shown within a large box, separated by a horizontal line.

pickEMs contains those electromagnets corresponding to the warp threads to be raised in a particular shed, allowing the brocade shuttle to weave a 'slice' through a character in the name tape. The particular subset of the *Electromagnets* that constitutes *pickEMs* is different in different picks. For example, Figure 8.8 shows the pattern of brocade threads for the letter 'R' (though with far fewer warp threads than is actually the case). The selection of electromagnets is a central function of *ControlBox*. Each character known to *ControlBox* can be woven in a sequence of picks, each pick weaving a slice of the character. Each shed pattern is stored on a punched card, and each character has a corresponding sequence of these cards associated with it.

Notice that there is a correspondence between the state changes of the cam contacts, electromagnets and solenoid, and the operations identified in the static model (turn-on, turn-off, etc). Notice also that the directions of the interaction arrows in the dynamic model correspond to those in the static model. Thus the cam contacts (the sensors) operate on the control box; the control box operates on the electromagnets and the solenoid (the actuators). The physical behaviour of the weaving frame (the sheds and the shuttle) is not modelled here, because we are primarily interested in the function of the control box. The behaviour of the cam contact is caused by the continuous driving of the cam shaft by the loom motor, but for our purposes the cam contacts can be regarded as generating signals spontaneously.

The progression of events is essentially the same as that described in the timing diagram. The optionality of the lifting box solenoid being switched on and off, as noted in the state transition diagram, is made explicit. There is a time constraint on *pickEMs*, pointed out by the loom operator: the electromagnets are not allowed to be on for more than 30% of a machine cycle, due to the danger of them burning out. We duly note this down. The sequence of interactions is as follows (the words in parentheses describe what is happening to the rest of the loom):

state 0 *cam1*, *cam2* and *cam3* are open (nothing is happening). *cb* receives a *nowClosed* message from *cam1*, and sends a *turnOn* message to *pickEMs*.

state 1 *cam1* is closed; *pickEMs* is on (and the jacquard is forming the correct shed, and fractionally later the shuttles start to be passed through the warp threads) *cb* receives a *nowOpen* message from *cam1*, and sends a *turnOff* message to *pickEMs*.

state 2 *cam1* is open again; *pickEMs* is off (the shed is still open, and the shuttle is still being passed through). *cb* receives a *nowClosed* message from *cam2*.

There is an option, depending on whether the next pick is to be a brocade pick or not.

'Brocade pick next' option:

cb sends a *turnOn* message to *lbs*.

state 3a *cam2* is closed; *lbs* is on (the shed is still open, and the shuttles are still being passed through; and the lifting box is starting to raise the correct shuttles into place for the next pick). *cb* receives a *nowOpen* message from *cam2*.

state 4a *cam2* is open again (the shed is still open, and the shuttles are still being passed through; and the lifting box is still raising the correct shuttles into place for the next pick). *cb* receives a *nowClosed* message from *cam3*, and sends a *turnOff* message to *lbs*.

'Ground pick next' option:

state 3b *cam2* is closed (the shed is still open, and the shuttles are still being passed through). *cb* receives a *nowOpen* message from *cam2*.

state 4b *cam2* is open again (the shed is still open, and the shuttles are still being passed through). *cb* receives a *nowClosed* message from *cam3*.

Back to normal:

state 5 *cam3* is closed; *lbs* is off (the shed shuts, the shuttles have been passed through, and the correct shuttles are in place for the next pick). *cb* receives a *nowOpen* message from *cam3*.

And so back to state 0.

The models we have produced so far do not mention the two most visible parts of the weaving process: the production of name tapes and the work of the operator. Feeling that we know more than enough at the moment about the operation of a loom, we next look at name tapes and the operator interface.

8.2.2 The name tape

The product of the weaving process, as suggested above, is ribbons of name tapes. Each brocade pick weaves one slice of a character, whether the character is one taken from the patch panel, or one such as the name separator, woven automatically by the control box when switching between patch panels.

Each name can be woven in either red or blue. A batch is of about 30 names, all in the same colour. Looking at a selection of ribbons, we discover the reason for the looms being *nearly* identical. Each name can be in one of two styles—large or small capital letters. Each control box, it turns out, is capable of weaving only one style, and since each loom is associated with its own control box, all name tapes for the 'large' style are woven on one loom, and all name tapes for the 'small' style are woven on the other.

8.2.3 The operator machine interface

The operator has two principal tasks: to set up the names to be woven at the interface with the control box, and to change the colour of the brocade shuttles.

The control box has four patch panels on its front, each of which has a matrix of holes, each hole corresponding to a letter (a capital letter in the style of the control box) and the letter's position in the text. The operator inserts pegs into these holes to represent the text of a name to be woven (the 'pummelling' activity seen earlier). The text is taken from a batch sheet, which specifies a batch number and up to 30 names. At any one time, text from one of the patch panels is being woven. Once the text is completely woven, the pegs are removed from this panel and the control box takes its instructions from the next panel nominated by the operator. The operator can now enter a new name into the free panel. This is a complex and error-prone interface, and, as we ascertained during preliminary analysis, is to be replaced, and so we do not pay any more attention to it.

The batch sheet also specifies in which colour the names are to be woven. Changing the colour of the brocade shuttles is quite separate from the control box, and involves delving into the weaving frame itself. This takes about an hour, and involves the operator, along with anyone else who is around to lend a hand. We decide not to investigate this activity any further, because we guess that this is not a process that will be different in the New World. However, we note that the onerous nature of colour changing is the reason why the batches are one colour, and why the operator shuffles through the sheets before picking one—to choose a batch of the same colour as is currently on the loom, and to assess when a colour change is needed.

We also decide not to pursue the packaging of name tapes any further, merely noting that the batch sheet is associated with the name tapes it specifies by being put into the same plastic bag and sealed. Over by the wall where the bags are stacked, there are two piles of batch sheets, one pile for each loom—since each loom weaves a certain style. The batch sheet originates in the order office, which is where we go next.

8.3 Orders, batches and the batching process

We know from our purposive modelling that this area, Orders & Invoicing, is where orders are received from customers and are restructured into batches of items to be woven by Manufacture. Thinking behaviourally, we might wonder what an order looked like, what a batch looked like, and how one is turned into the other.

Our knowledge of what an order looks like comes from looking at a selection of them, and talking to the order office staff. Orders come from customers and specify the customer's address and a number of order items. An order item specifies a name to be woven, the quantity required, and the style and colour of the name tape text. There are two kinds of order: those that come from a pre-paid book, specifying only one order item, and those that come from magazine adverts or a written request, accompanied by cash or a request for an invoice. However, there is no difference in the way that these are treated for the purpose of manufacturing.

A batch is a collection of up to 30 order items, all of which, where possible, have the same colour and the same style. For each order item, its text and its order number are transcribed on to a batch sheet, which is passed on to the factory floor.

We can model this in Beluga as a class *Order* with features holding various customer details (Figure 8.9). Each *Order* is associated with one or more *OrderItems*, and each *OrderItem* is eventually associated with a *Batch*. A *Batch* may contain up to 30 order items.

The batching process itself is extremely complex. It involves juggling order items between 'proto-batches' in a difficult-to-define manner, until the batches

8.3 Orders, batches and the batching process

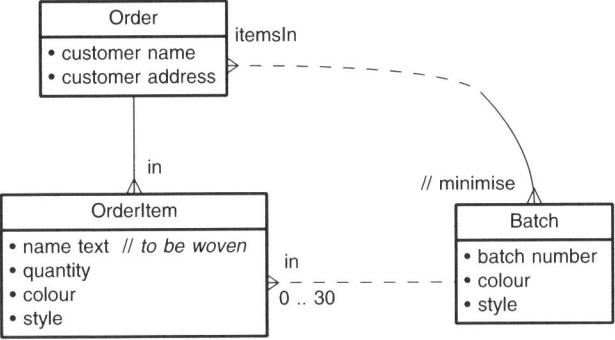

Figure 8.9 A static view of orders and batches

satisfy certain desirable characteristics to some difficult-to-define degree. So, it is easier to describe the process simply in terms of the desirable characteristics. Three of these are described in the model above:

- Each batch contains order items sharing a single colour and a single style.
- Each batch contains up to 30 order items.
- The number of batches to which order items from a single order are allocated is kept to a minimum. Usually, the minimum possible is four, because there are four colour/style combinations. Occasionally, order items from a multiple-item order have fewer than four colour/style combinations.

The first two of these are derived from the needs of the weaving process; the third is derived from the need to associate name tapes with their order. We can note, as a result, that the characteristics of a batch are likely to change if there is any significant change to the weaving process.

We can describe the overall behaviour of orders, order items, and batches as two concurrent 'processes'. The first of these processes takes in new orders and order items within these, and then waits until all items in an order have been woven, before 'signing off' the order. The second process groups order items into batches (perhaps in an incremental fashion) and then weaves the items in batches—which is where the loom becomes involved.

This is shown in Figure 8.10 with the first process drawn above the second. *OneOrder* shows the behaviour of a stereotypical order. The order is received, and given *pending* status; its order items are initially *unbatched*. In effect, a new order and set of order items is created (although the details arrive by post); this is indicated by the blob-with-dot icons. At some later time, all the order's items have been woven, and the order itself is given *woven* status. Since the order items may well be in different batches, they achieve *woven* status at different times. A simple annotation to the model records the requirement that turnaround time for one order, and so for its set of order items (between being received and being

84   Chapter 8   Old World Behaviour

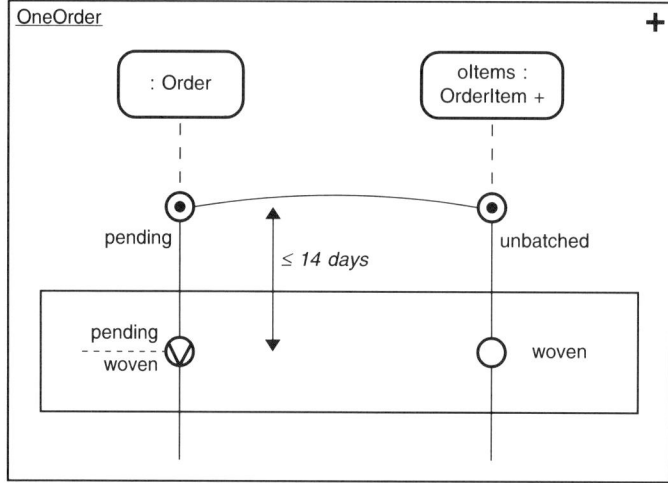

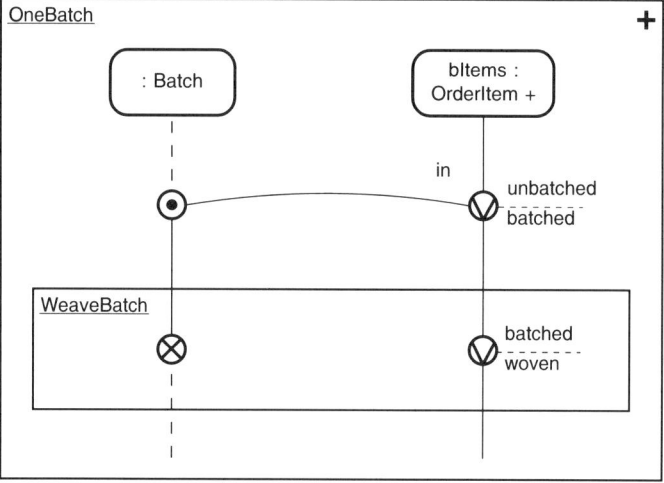

Figure 8.10 A dynamic view of orders and batches

woven) is 14 days or less. The '+' at the top right corner of the *OneOrder* box indicates that the process deals with multiple orders, and that these are handled concurrently (that is, many orders can be in existence at the same time).

Independently, the second process is creating and processing batches. *OneBatch* shows the behaviour of a stereotypical batch. Once a new batch has been created, a set of unbatched order items is put into it, becoming *batched*. This may happen 'in one go' or incrementally, a few items at a time. At some later time, the set of batch items is woven, finishing the batch. This behaviour is called *WeaveBatch*. The '+' at the top right corner of the *OneBatch* box indicates that the process deals with multiple batches, and that these are handled concurrently (that is, many batches can be in existence at the same time).

The link between the two processes is that every order item eventually appears in a batch. The turnaround time constraint for order also links these two behaviours: the order items from one order all have to progress through the second process, as batch items, within the 14 day slot from when they were received within an order.

8.4 From orders to name tapes

We have investigated behaviour in five areas of NIMWeC, all related in some way to the production of name tapes to order:

1. The weaving process—produces woven ribbons, according to specified patterns.
2. The name tape—a woven ribbon with various names on it, corresponding to the items in a batch.
3. The operator-machine interface—takes details of a batch and feeds them into the machine, item by item.
4. Orders—contains order items, which hold details of what is to be woven.
5. Batching—takes order items from orders and groups them into batches according to colour/style constraints.

We might wonder if this tells us all we need to know about how an order becomes woven, and the final product delivered. A typical scenario of an order proceeding through NIMWeC is as follows:

- An order arrives, and is broken down into its items (area 4, above).
- These items are put into batches (5) and the details are passed to Manufacture.
- Batch specifications are taken one at a time by a loom operator and the details of the batch items are input via the patch panel (3).
- The loom weaves batches and items within batches as a sequential process (1).
- The woven ribbons (2) are cut up and sent back to the order office where the woven items from the ribbons are matched up with the originating order items (and order).
- Once all the order items are accounted for, the name tapes are sent to the address specified on the order.

We can now link together the behaviour modelled so far to reflect this 'big picture'. The 'missing link' is the idea of an *Instruction*. Each order item specifies a name to be woven, which consists of a sequence of characters. For every such sequence of characters (and appropriate separators) there is a sequence of instructions, each

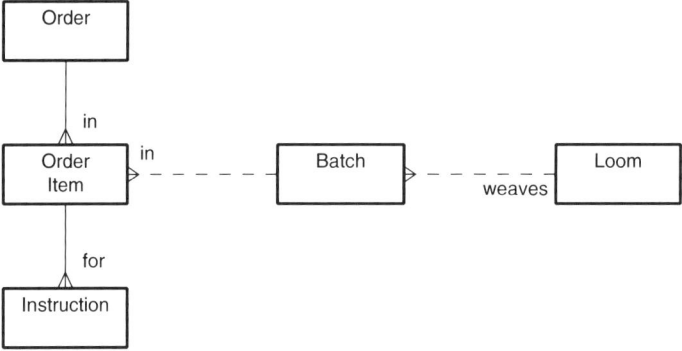

Figure 8.11 A static view of weaving a batch

instruction determining what the loom does for a slice of the name tape. We know that each batch is assigned to a particular loom (on the basis of style/colour). If we then assume that items are ordered within batches, and batches are dealt with in some order, it follows that we can generate a single sequence of instructions to control the loom, and so weave the required name-tape ribbons. A static view of this is shown in Figure 8.11.

Looking at the dynamics, we have three levels of 'nesting' (Figure 8.12):

- *OnePick* machine cycle and *Instruction*, within *WeaveItem*
- *WeaveItem* within *WeaveBatch*
- sequence of *WeaveBatch* behaviours

One *Instruction* controls one *OnePick* behaviour (see Section 8.2.1), by:

- determining which electromagnets are in *pickEMs*
- determining whether the next cycle is a ground pick or a brocade pick

Exactly how this information is represented within an instruction, and how it is interpreted is not modelled here, since we feel that this is an area that may well be redesigned for the New World. We regard the instruction as being 'consumed' (indicated by the blob-with-cross icon) since it has no further involvement in the behaviour.

Once all the instructions for an order item (in a batch) have been processed, the item becomes *woven* (the dependency is shown by a double-headed arrow). This is the change of status that is of interest to the order-handling process (see Section 8.3). The physical constraint that the machine cycles happen sequentially is shown by the inner double-sided box.

Once all the order items in a batch have been woven, the batch is finished (and has no further involvement in the behaviour). The physical constraint that the batches are woven sequentially is shown by the outer double-sided box.

We now have a more detailed description of the *WeaveBatch* behaviour that we identified in Section 8.3. This links the *OnePick* behaviour describing a single

8.4 From orders to name tapes 87

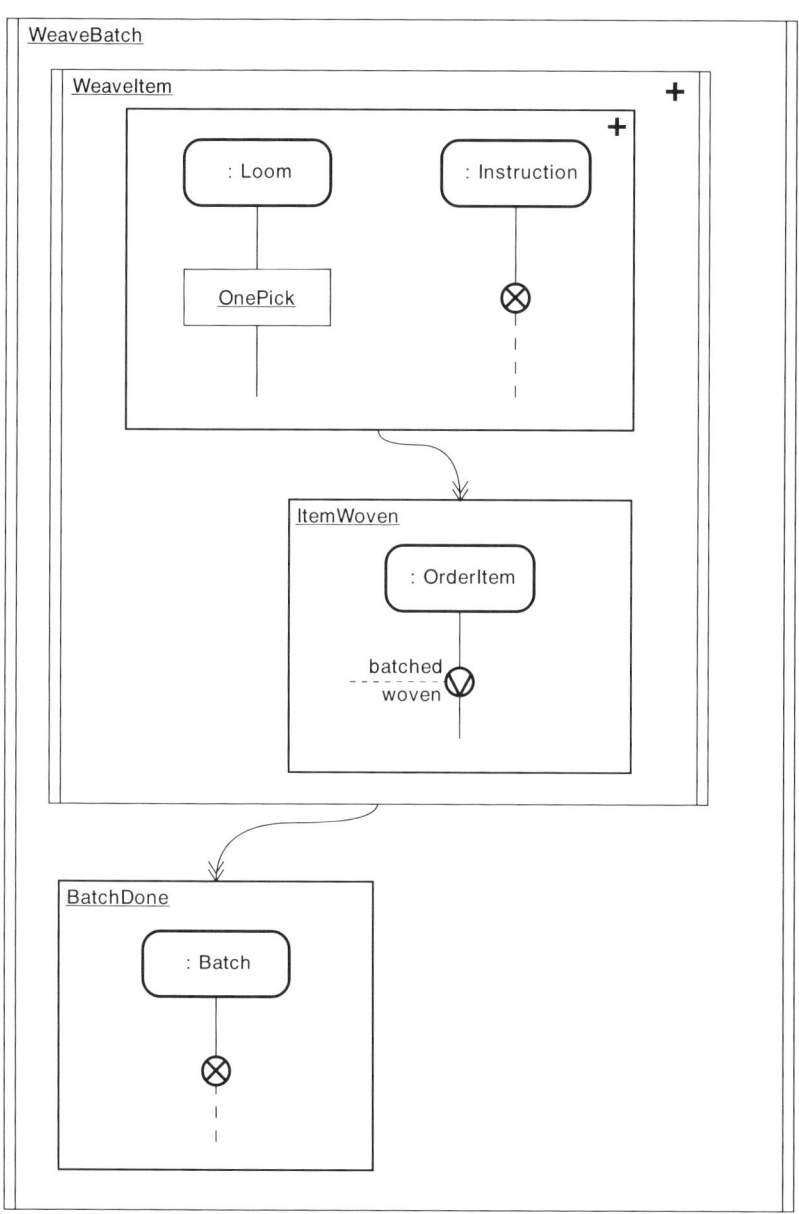

Figure 8.12 A dynamic view of weaving a batch

machine cycle for the loom, with the processing of customer orders described in Section 8.3.

In our analysis of NIMWeC, we now feel that we have a fairly sound understanding of how the present system works. In particular, it has become clear to us what behaviour realises the co-operations between Manufacture and Orders & Invoicing. We have also gained a good grasp of the rather arcane terminology used by our client. We now feel confident in progressing to the next activities of the analysis: diagnosing a pathology, and assessing alternative prescriptions for remedying the pathology.

Chapter 9
Pathology and Prescriptions

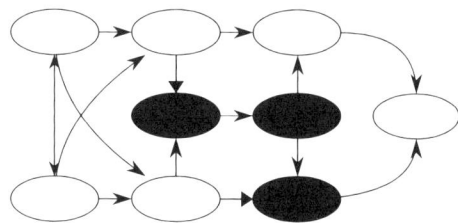

9.1 Introduction

A *pathology* is an explanation of the problems with the Old World in terms of underlying causes. The problems might be with the way the world currently works, or that the world is unable to change in certain desired ways. A *prescription* is a remedy for a pathological situation.

Chapters 7 and 8 build a model of parts of the Old World purpose and behaviour. This model can be altered by 'adding in' the immediate effects of the development objectives, in terms of what the new purposes of the various parts of NIMWeC look like.

The pathology/prescription activity involves tracing the consequences of these changes to the models. The aim is to discover which parts of NIMWeC are unsuitable, or pathological, with regard to the new demands placed on them. Current pathologies within NIMWeC are also noted. A prescription can then be posited. This typically prescribes some changes to behaviour, that remove, or go some way towards removing, one or more pathologies.

Some combination of compatible prescriptions together constitute the course of action that should be followed in order to result in a satisfactory New World NIMWeC. In order to choose a suitable bundle of prescriptions, an evaluation of the costs and benefits of each has to be made.

As with the design and execution stages of the process, and the activities of the Basic Process, progress within this activity is unlikely to be linear. Each prescribed change to a model may generate further pathologies, requiring new prescriptions, and so on. However, it is presented below as a linear process, with the aim of giving a clear picture of pathologies and prescriptions themselves.

9.2 Adding in the development objectives

There are four main development objectives identified in the preliminary analysis. The client wants the New World NIMWeC to achieve the following:

1. sell twice as many name tapes
2. make fewer errors in the weaving of name tapes
3. be able to sell into foreign markets (in particular, Sweden and Israel)
4. produce statistics about what styles and colours are being sold, and when, to allow knowledge about current demand and prediction of future demand

Identifying how these objectives relate to the models that we have produced is reasonably straightforward.

The first of these involves a change to the top level NIMWeC guarantee to 'supply a range of woven labels in a timely manner'. This is amended to 'supply a range of woven labels in a timely manner, at double the Old World volume of sales'.

The 'errors' mentioned in the second objective are behavioural errors in the transfer of a weaving instruction to the machine, which results in name tapes being woven that do not match the instruction. The problem area is the interface between the Operator and the Machine—the peg-board patch panel of the control box. This objective therefore concerns a specific bit of unsatisfactory behaviour, which is to be improved in the New World. There are no further implications for the Old World.

The third objective raises many questions:

1. Is there a suitable customer base? Do the Swedish want name tapes, whether for school uniforms or not?
2. Can Sales & Marketing create and sustain a suitable customer base? Do they know the best way to advertise in Israel? Can they find new uses for name tapes in foreign markets?
3. Will the demand be such that further increases in production capacity are required?
4. Is the product range suitable? Should the name tapes be different in any way for Sweden or Israel—such as being of a mandatory size or colour? Are Swedish and Israeli names weavable?
5. How would ordering, payment and dispatch be handled? Would dealing directly with customers be feasible, with postage to UK and payment in UK Sterling? Can Orders & Invoicing read Swedish? Could suitable outlets or agents be found, and could these handle the currency and postage issues?

It transpires that the client wants recommendations on the feasibility of selling into foreign markets, rather than a specification that allows this to happen now, and so we note most of the questions as issues that would need to be addressed. However, the third and fourth questions do seem relevant to the current project.

If the client currently wants to double production for the UK market, then it is likely that a further increase in production will be necessary to cope with selling into foreign markets as well. The part of the model that is affected is the same as for the first objective—the top level NIMWeC guarantee to supply woven labels.

Is there anything fundamental to name tapes that would need to be different? Some consideration of typical Hebrew names suggests that a fundamental problem with selling into foreign markets will be the alphabet of the foreign language. We know from the behavioural modelling that the problem with the current world lies with the control boxes, each of which can only offer one style, in capital letters. As with the 'remove errors' objective, the part of the model affected by this development objective is a specific piece of behaviour, and so we make no change to the model as yet.

Reflecting the fourth objective in the models involves adding a new reliance, on statistics, to Sales & Marketing acting as the representative of our client's forward planning activities.

9.3 Structuring pathologies, positing prescriptions

The starting point for considering pathologies is the change to the Old World that is needed to reflect the development objectives. It is simplest to take each development objective in turn, and then trace through the consequences of the associated changes to the model. Prescriptions are posited for any resulting pathologies.

9.3.1 Sell twice as many name tapes

Pathologies

We start by identifying any co-operations and delegations leading out from the guarantee 'supply a range of woven labels in a timely manner, at double the Old World volume of sales' on NIMWeC. Since this guarantee has been changed (in order to reflect one of the development objectives) the co-operations and delegations effectively become unanalysed once more, and may be problematic. The trace of co-operations and delegations goes out from NIMWeC to Customer Base, and in to Manufacturing and Orders & Invoicing (Figure 9.1). Not all of these co-operations and delegations are problematic. The client has told us that the customer base, through both magazine and pre-paid orders, can provide demand for doubled production, and so the co-operation between NIMWeC and Customer Base remains unproblematic. We assume from now on that any increase in production capacity will result in an increase in orders and actual production. The client also tells us that Orders & Invoicing can cope easily with the extra payments.

The other delegations to Orders & Invoicing, and the delegation to Manufacturing are both problematic, however.

92   Chapter 9   Pathology and Prescriptions

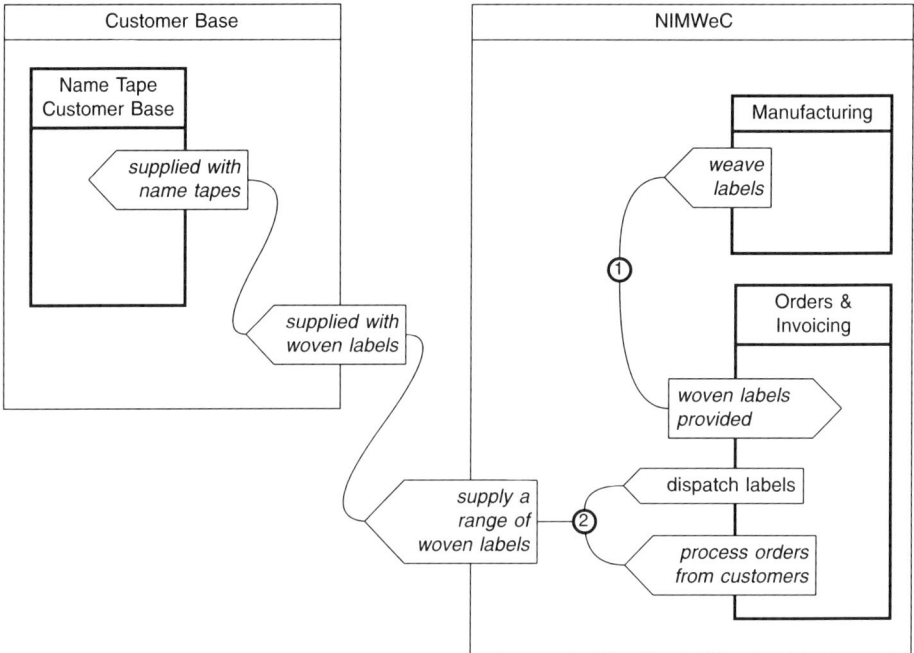

- qualifiers as in Chapter 7, except for:
- NIMWeC: *supply a range of woven labels* in a timely manner, and at double the Old World volume of sales

Figure 9.1 Tracing NIMWeC co-operations

The problem with Orders & Invoicing 'bottoms out' at the guarantee on Name Tape Orders & Invoicing to 'process orders from customers'. The behaviour underlying these involves batching, and the transcription process from order form to batch sheet. All of this takes time to do accurately. Neither activity could keep up with the demands of doubled production.

The problem with Manufacturing 'bottoms out' in two places: at Looms' guarantee to 'weave labels', and at the co-operation between Operators and Looms (Figure 9.2). The first of these relates to the behaviour of the looms and their control boxes. It can be viewed as a problem with the loom's control boxes: they break down frequently, with a characteristic grinding noise, and so are the cause of a significant amount of each loom's downtime. It can also be viewed as a problem with the inherent capacity of the current two looms, which, when their control boxes allow, are running somewhere close to maximum capacity.

Taking the second of these, there are two aspects to the co-operation between Operator and Looms.

One aspect is the need for a loom's brocade shuttles to have the correct colour of weft thread for the batch it is about to weave. Our client tells us that the changing of the brocade shuttles' threads cannot itself be speeded up, for example

9.3 Structuring pathologies, positing prescriptions

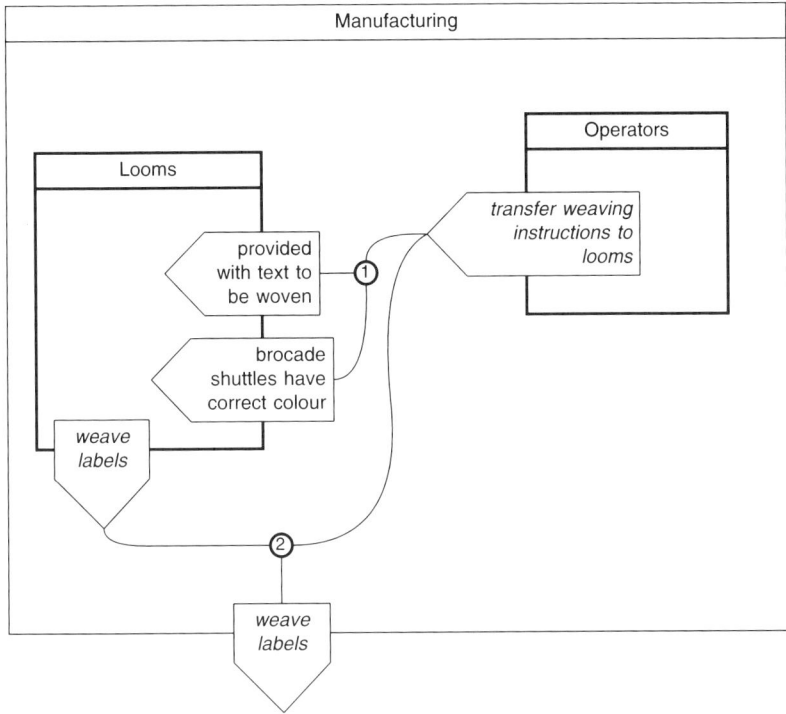

- qualifiers as in Chapter 7

Figure 9.2 Tracing deeper into NIMWeC

by more staff working on it at once. The client also tells us that the expense and complication of trying to reduce the need for thread changes through mechanical means, such as using larger shuttles with more thread, or inventing an 'endless thread' shuttle mechanism, would outweigh the likely benefits. The aim in the past has been simply to try to avoid, through the sequencing of batches, the need for colour changes other than when the shuttles have run out of thread.

The other aspect of the co-operation manifests itself in the patch panel attached to the control box, where the operator enters the text of batches to be woven. It is unlikely that the operator could keep up with doubled production using this interface, due to the time it takes to set up each patch panel accurately.

Prescriptions

We can now look at each of these basic problems and identify potential prescriptions for each. At this point there is a shift in our perspective. In building up the models and pathologies the source of our information was the organisation, its owner, its staff and our observations in the factory. We used some generic models that we as analysts brought with us to the analysis, in order to help us to organise our understanding of that information. In order to generate prescriptions we call on information from other sources: our experience of similar projects, and our knowledge of available specifications and products. This is analogous to the situation of a doctor who, in order to cure an illness, rather than merely diagnose it, needs knowledge of pharmacology, physiotherapy, and all the other means of restoring the patient to good health.

For NIMWeC, we are guided mainly by previous experience, and by what was made clear during the preliminary analysis about the way that the client envisages the New World—for example we do not consider the prescription 'move site to take advantage of favourable tax arrangements'. However, we try hard not to be seduced by any prescription that seems to be the obvious choice. Any prescriptions that are in any way unacceptable are weeded out during the course of evaluation (as described in Section 9.4). The prescriptions, being quite simple, are expressed in plain text. If it were important to be more precise at this stage, then they could be expressed through models of the prescribed purpose or behaviour (which we do later when evaluating the prescriptions).

Candidate prescriptions for the transcription and batching process being unable to cope with doubled production are:

1. Do nothing.
2. Provide better lighting in the Orders & Invoicing office.
3. Provide an automated order-reader, removing the need for transcription altogether.
4. Standardise order forms, and pass these directly to the factory floor, removing the need for transcription altogether.
5. Simplify the batching process, perhaps by offering only one colour of name tape.
6. Provide an automated batcher to take individual orders and produce optimised batches of order items.
7. Increase the number of personnel in the Orders & Invoicing office.

Doing nothing is always a possible prescription, albeit one that is unlikely to achieve the development objective.

Candidate prescriptions for the Operator Machine interface being unable to cope with doubled production are:

1. Do nothing.
2. Change the interface to the control box by replacing the patch panels, to allow input to keep up with production, by means of easier input and/or better feedback and/or an ability to enter weaving instructions more than four names ahead at a time.
3. Send the operator on a dexterity training course.

Candidate prescriptions for Manufacturing having only half the desired production capacity due to the unreliability of the control boxes are:

1. Do nothing.
2. Improve the reliability of the control boxes through the application of copious quantities of lubricant.
3. Replace the control boxes with one or more new and more reliable control systems.

Candidate prescriptions for Manufacturing having only half the desired production capacity due to the inherent capacity of the existing two looms are:

1. Do nothing.
2. Speed up the looms through the application of copious quantities of lubricant and the introduction of PTFE-coated kevlar super-shuttles.
3. Increase the number of looms to 3 or 4, using the 2 spare jacquards that can be freed from the existing looms. This entails providing a suitable control system for the extra looms; the client has said quite specifically that buying more of the current control boxes would be inappropriate.

9.3.2 Make fewer errors in the weaving of name tapes

Pathologies

The starting point here is the behavioural Operator Machine interface. As a result, there is no trace for this development objective, just a note that it is the behaviour underlying the interface (the patch panel again) that is the problem.

Prescriptions

The candidate prescriptions are broadly the same as those given for the Operator Machine interface in the previous section, only here they are prescriptions for this interface being error prone. They are:

1. Do nothing.
2. Change the interface to the loom to allow more accurate input, by means of easier input and/or better feedback.
3. Send the operator on a dexterity training course.
4. Have two operators per loom rather than one, to reduce pressure on individuals.

9.3.3 Sell into foreign (in particular Swedish and Israeli) markets

Pathologies

The changes to the Old World involve an increase in production over and above any currently proposed increase, and a change to the styles (specifically, characters) that can be woven. The parts of the models affected for the increase in production are the same as those for doubling production, and the trace for this problem would be the same as for that one. The part of the model affected by the need to weave different characters is just the control box—and so there is no trace for this.

Prescriptions

Candidate prescriptions for the need to increase production in the future are:

1. Do nothing.
2. Provide an easy route to more than doubling production through any of the measures proposed for initially doubling production.

Candidate prescriptions for the need to allow different characters to be woven are:

1. Do nothing.
2. Replace the control boxes with something or things that can weave foreign characters.

9.3.4 Produce statistics about what is being sold, and when

Pathologies

The change to the Old World involves adding a new reliance to Sales & Marketing: 'provide statistics about what styles and colours are being sold, and when, to allow knowledge about current demand and prediction of future demand'. This is simply an unfulfilled reliance, and so, as with the 'make fewer errors' development objective, there is no trace of the consequences of this—any further changes to the model will be manifestations of specific prescriptions.

Prescriptions

The candidate prescriptions are again simple, and so expressed in plain text. Prescriptions for Sales & Marketing having an unfulfilled reliance on statistics are:

1. Do nothing
2. Get Orders & Invoicing to produce information, if possible in a way that won't involve too much new behaviour, provide suitable behaviour in Sales & Marketing to support the bulk of the co-operation, and amend current behaviour in order to do useful work with the information.
3. Get Manufacturing to produce information, if possible in a way that won't involve too much new behaviour, provide suitable behaviour in Sales & Marketing to support the bulk of the co-operation, and amend current behaviour in order to do useful work with the information.

Implementing any of these prescriptions will involve a change to the model—adding a reliance to Manufacturing, for example. What should actually happen is that this becomes the starting point of another pathology/prescription process. For now we just note that either Orders & Invoicing or Manufacturing will have to be called on to produce information about orders.

9.4 Evaluating the prescriptions

So far we have gained an understanding of the current situation, identified the consequences of the development objectives and how these make the current situation unsatisfactory, and produced prescriptions that describe candidate solutions for one or more of the pathologies. Now we have to consider how the prescriptions can be 'bundled' together. The resulting candidate bundles can be evaluated, and one or more chosen to be implemented. The New World can then be specified according to the chosen bundle (see Chapter 10).

Each prescription is analogous to a hypothetical development objective at a greater level of detail. As we did with the development objectives, we need to assess the consequences generated by the implementation of each prescription. We do this by following a similar pathology/prescription process: we posit the prescription as a change to a role in the Old World; we identify the roles that are affected by this change; we analyse the effect to determine if it results in a pathology; we identify prescriptions for any pathologies. The additional task involved here is in deciding where to start.

9.4.1 Choosing a starting point

In inventing prescriptions, we ruled out any obviously unacceptable ones such as 'move site to take advantage of favourable tax arrangements', before they were even written down. Even allowing for this initial pruning, for an analysis of any size there are too many prescriptions simply to assess the possible effects of all the individual prescriptions and all the possible combinations of prescriptions.

This a general problem with the engineering endeavour of finding suitable solutions to complex problems. Some resources for dealing with this, as in the initial generation of prescriptions, are our experience of similar projects and our knowledge of available specifications and products. In addition, common sense suggests that focusing our attention on the central problems is likely to lead us to a satisfactory solution, if not necessarily the optimal one.

If we are to double production, then the obvious central problem is Manufacturing having only half the desired production capacity, and so we concentrate on this. We identified two problem areas with regard to this: the unreliability of the control boxes, and the inherent capacity of the looms.

Anecdotal evidence from the operators suggests that downtime attributable to the control boxes is the cause of the looms running at about 80 per cent of potential capacity. Even a complete improvement in the control boxes' reliability, by means of either of the prescriptions that we have posited (the application of lubricant, and the complete replacement of the control boxes), would not result in the desired production level—so not much benefit, albeit at not much cost.

The second problem area is that of the inherent capacity of the two looms. The first prescription involves lubricant and lightweight low-friction components in order to speed up the running of the looms. The increase in speed that is likely through such efforts, we decide, would be minimal. The second prescription involves buying more looms, and does hold the possibility of significantly increasing production. We decide to focus our attention on this prescription, bearing in mind that this would be likely to form a component in an expensive bundle of prescriptions, and that we might want to consider what a less expensive, if less effective bundle might be.

9.4.2 A first candidate bundle

The trace for this prescription starts with buying more looms, and the appropriate rearrangement of the Jacquards. We are assuming that a new loom comes complete with Lifting Boxes and Camshafts.

The first thing we have to consider in our trace is the collection of electromechanical devices connected to the loom: the Electromagnets, Lifting Box Solenoids, and Cam Contacts. We need a new set of these for each new loom.

In addition, we need a way of controlling the new looms. There are several choices here. We could buy more of the current control boxes, but we have been told quite strongly that the client does not want to purchase any more of them. This leaves us with a choice between different kinds of new control systems:

- a new control system that mimics the old control box, but in a reliable medium, for each new loom
- such a new control system for each of the looms, both old and new, and the disposal of those control boxes that we have already
- a central control system that controls all of the looms

In each case, we could reverse engineer the old control box, and provide a replacement that perhaps worked not just more reliably, but better, for example, by allowing all looms to weave all styles.

If we change the control boxes at all, we need to consider the Operator Machine interface.

Firstly, with an individual control system for each loom, we might need more operators, although the control systems might work efficiently enough to allow an operator to manage two looms; we might need fewer operators with a centralised control system.

Secondly, we need to consider the interface to the new control system. Preserving the current interface would not be a good idea.

Thirdly, we may want to change the nature of what the operator transfers to the machine. Is a batch still going to be the same? Can more names now be in a batch? Do we need batches at all, or could the operators cope with a continuous string of order items? Are there any other constraints on batch that might change?

Finally, we can complete the trace in our model, from Batch to the batching process. If a batch is no longer the same, then the batching process will probably have to change. The number of batches required is increased significantly, and so we have to address the problem of the increase in effort that is needed for the transcription and batching of orders, offset against the likelihood of the batching process itself becoming simpler. Any of the more effective prescriptions for the transcription and batching process problem would be suitable here, the simplest being to increase the number of personnel in the orders office, if necessary.

Another complication associated with batching is that previously it was clear which loom would weave which batch. With several looms capable of weaving all batches, there is some work involved in allocating batches to looms. Achieving an optimal allocation, with regard to minimising colour changes on looms, is potentially a complex process. This could either be undertaken manually or could be dealt with by the control system.

Looking back to our original list of atomic prescriptions, we can see that several of these are encompassed within the above trace, and those that are not subsumed or made unnecessary do not conflict either. In fact, it turns out that the consequences of this prescription—buying more looms—encompasses a prescription that addresses most of the problems that we had identified. This should not be too surprising, since two of these problems were perceived in response to doubling production (transcription and batching being unable to cope with doubled production, and the Operator Machine interface being unable to cope with doubled production), one is subsumed altogether (the above prescription addressing doubled production through addressing control box unreliability) and one addresses a common system (the Operator Machine interface being error prone).

The problem of allowing for greater expansion (for future expansion into foreign markets) would be catered for by choosing a control system that could deal with any number of new looms. The problem of having a restricted character set could be solved by having some or all of the looms with a control system that allowed for foreign characters.

Some rudimentary statistics about orders (number, details, dates) could be provided by a new control system, although this has to be with the third option above, a central control system, for the statistics to be complete. More wide-ranging statistics involving customer details (geographic distribution of customer base? favoured kind of payment? number of payments outstanding?) is beyond the scope of this prescription, however. The question of what statistics are needed, how and how often the data is to be obtained, and how the statistics are to be calculated and presented, is also an issue that lies outside the scope of this prescription, beyond noting that the control system would have to be able to provide data about orders, frequencies, and so on.

So, this one prescription, based around a centralised control system for both old and new looms, has the potential of addressing almost all of our problems. This seems likely to be an acceptable solution.

Had this prescription not addressed most of our problems, or not had the potential for doing so satisfactorily, then we would go on to investigate other prescriptions. We would choose a prescription to look at by using the same approach of finding a problem that appears to be central, and investigating a prescription addressing that problem that survives the scrutiny of a rudimentary cost/benefit analysis.

9.4.3 An alternative bundle

Although this prescription does seem suitable, we would like to present our client with some alternatives with regard to the size of the development being proposed. The prescription that would be central to an alternative low-cost bundle is the one briefly considered above, of increasing the capacity of the two looms through the application of lubricant, which would make the current Control Boxes as reliable as possible.

The trace for this prescription starts with a slight increase in the production capacity of the Looms and control system. The trace can be followed to the operator machine interface. Can the operator now keep up, in order to make use of the increase in capacity? A slight increase in the efficiency and accuracy of the interface would be required—perhaps some feedback, or a dexterity training course for the operator. If the operator can be made to keep up, then the next question is whether the orders office can supply batches quickly enough. Again, a slight increase in capacity would be required. As noted in the preliminary analysis, the factory is rather dark—perhaps just better lighting would speed up the transcription of orders sufficiently.

Again, several of the other atomic prescriptions are encompassed within the above trace, and, given the partial nature of these prescriptions, none conflict. However, the prescription goes no way towards either expanding into foreign markets or to providing useful statistics.

An alternative higher-cost bundle, perhaps with a more extensive ordering and statistics capability, is unlikely to be acceptable to the client, and so we stop at two bundles.

9.4.4 Evaluating and choosing

Having built up candidate bundles of prescriptions, the cost, benefit, risk, feasibility, and so on of each needs to be determined in order to allow for a reasoned decision to be made about which to choose. Different options meet development objectives to different degrees, and it is up to the client to choose one. We also have to evaluate each option with regard for the feasibility of satisfactorily achieving an implementation, and the associated cost of the migration from old systems and work practices to new ones.

For NIMWeC, we have identified two such bundles above. As with prescriptions, do nothing is still an option, so we have three bundles to evaluate: do nothing, do a little, and do a lot.

Do nothing

There is no immediate cost associated with doing nothing, since there is no pressing need for NIMWeC to change. There is also no benefit, other than being able to use elsewhere (or not use at all) the funds earmarked for a future development resulting from this project. An evaluation of the long-term cost of doing nothing is outside the scope of this project.

Do a little

The bundle of minor improvements, involving improving the control boxes' reliability, and making corresponding changes to other parts of NIMWeC, would cost very little in monetary terms, be easy to implement, and would have little effect on the profitability of NIMWeC. The control boxes' unreliability currently limits the capacity of each loom to about 80 per cent of potential capacity. If the reliability of the control boxes were improved, and some improvements made to the interface, then primarily through better reliability, and partly through increased accuracy of operator input, we might expect each loom to work at near full capacity, resulting in a total increase of capacity over the Old World of about 25 per cent. The resulting changes in working conditions, such as better lighting, might be welcomed. They might also improve the ease with which NIMWeC satisfies current health and safety regulations.

Do a lot

If we bought another loom and utilised an excess jacquard freed from one of the existing looms (there are currently four jacquards, but only one is needed per loom), and had all the looms using a new control system, then we could expect a total increase in capacity of about 90 per cent—nearly double. Bringing the total number of looms up to four would result in an increase in capacity of about 150 per cent—more than double.

With a centralised control system there should be the possibility for future expansion, along with the ability to cater for differing market demands for name tapes, or indeed general woven products.

The major cost will be for the new looms and the control system, with some cost associated with time taken for installation, testing and training. No new loom operators are envisaged, the simplicity of the new control system outweighing the number of extra looms, but there may be a need for extra maintenance effort, although, again, the expected good reliability of the control system may mean that there is enough maintenance effort available currently. The extra cost of providing a control system that also dealt with batch allocation would not be significant.

An installation starting by using the control system for one new loom and then progressively adding in new and existing looms, would be a sensible approach, both from the point of view of reducing risk and spreading out the cost over a period of time. It would, however, involve some extra cost (in 'missed opportunity') over the option of an immediate change-over to the new system.

Other costs associated with the new system include extra effort in the ordering department, since any savings from the potential simplicity of the batching process would probably be outweighed by the increase in effort on the greater number of orders. Some extra staff may be needed if indeed the extra production capacity were utilised.

We had already ascertained, during preliminary analysis, that the client is convinced of the need to replace the control boxes, because of their unreliability, unpleasant operator interface and the limits they place on future expansion. Not surprisingly, the client chooses the bundle that contains the replacement of the control boxes as one of its prescriptions.

Chapter 10
Specifying the New World

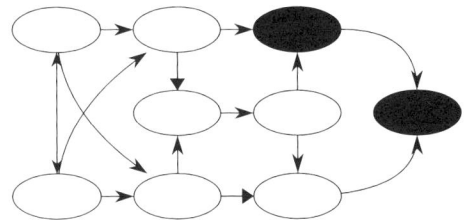

10.1 Introduction

The bundles of prescriptions discussed in Chapter 9 outlined courses of action that involve changes to part of the world. In general, there are various kinds of change that a prescription can describe:

- improving things (improving lighting in the ordering office)
- replacing things (replacing the control boxes)
- adding things (adding a statistics system)

Where a change involves the development of an IT system (involving both software and hardware) we can be more specific about each kind of change:

- providing an IT system that enhances or provides a better interface to an existing component
- replacing a manual or mechanical component with an IT system (for example, replacing the mechanical control box with a new computerised control system)
- adding an IT system to do new work

The bundle we have chosen for NIMWeC involves elements of all of these kinds of change.

In this chapter, we provide enough of a specification of the changes to allow suitable IT systems to be designed and implemented—the task of system development. The specification should form the basis of subsequent development work.

We are interested in particular in the IT systems that are being developed—we were hired on the basis of being IT analysts, and our knowledge of IT has had an influence on the nature of the prescriptions we produced. However, we are not yet at the stage of being interested solely in the IT systems themselves. We still need to produce a specification of the non-IT changes, such as changes in personnel or work practices. For the IT changes, we need to produce a specification not in terms of the internals of the systems, but in terms of what the purposes of the systems are, and how they are expected to behave with regard to the systems around them.

10.2 What the New World looks like

Many things in the New World remain as they were in the Old World. The purposes of NIMWeC remain broadly the same, but need to take account of the development objectives. Dispatch and payment, sales and marketing, loom maintenance and the manual procedures for changing the brocade colour of looms all remain as they were. These can cope well, we understand, with doubled production. Turnaround time, NIMWeC's big competitive advantage, remains at 14 days.

Each loom still expects something to give it instructions. We still want to be able to produce a batch from the orders office, each of a manageable 30-name size, and pass these down to the factory floor.

The main behavioural changes are outlined in the chosen prescription and centre on the weaving process:

1. The control boxes are replaced by a central control system for the old looms, with two new looms arriving quite soon.

2. The batching process is simplified, with all of a batch's items now simply being all of one colour, rather than all of one colour and one style.

3. The product has the potential for being different—not simply capital-letter name tapes, but ribbons with the Hebraic alphabet, or any pattern at all (within the limit of the resolution afforded by the looms).

The second and third of these changes have consequences for the first, as demonstrated in the pathology/prescription activity. The new control system has to deal with batch allocation, has to offer several styles, and has to ensure that it is easy to define new ones.

Much of the specification of the New World can be derived from our models of the Old World, since the latter was modelled at a suitable level of abstraction, deliberately ignoring details that we felt were liable to change. Our general approach is to view the control system as central, and describe other things around it.

10.2.1 Processing batches

The new *Batch* is similar to that described in Section 8.3, but no longer has a *style* attribute (Figure 10.1). The batch is the basic unit that the control system receives from the operator. Orders and batches still have the behaviour described in Section 8.3, and the timing constraints expressed there still apply. The orders office must be able to produce twice as many batches, and an extra person there may be needed to help with this; the nature of the jobs of those already there are

10.2 What the New World looks like

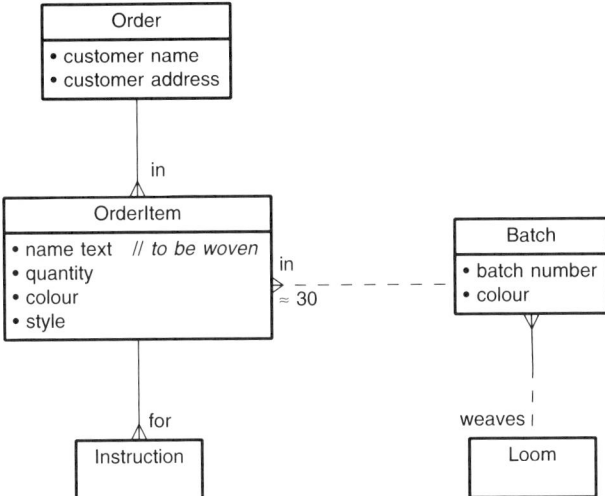

Figure 10.1 Batches in the New World

changed slightly, with less effort being spent on the batching process, some of this envisaged spare effort being spent handling the extra volume of orders.

At the other side of the control system, we now have several looms, expecting to receive instructions according to the *OnePick* behaviour of Section 8.2.1. However, as noted above, the bundle of prescriptions involves two major changes to the weaving process. One is that the one control system is to service all looms. The other is that the nature of the instructions that the looms expect has changed.

Since, in the Old World, each batch is automatically assigned to one loom, the relationship between batch and loom is made immediately. In the New World the relationship between batch and loom has to be established by the control system— the allocation problem. So, the first task of the control system is:

- to allocate each batch to a loom, in some fair way, on the basis of the colour that each loom is weaving, and the number of batches already allocated to each loom

The control system should probably not allocate batches too far in advance, due to the likelihood of loom breakdown disrupting an otherwise perfect schedule. There are many ways of dealing with the allocation task, but for this specification we simply note that at some point between a batch being created and the *WeaveBatch* behaviour the batch has to be allocated to a loom (Figure 10.2). Compare this diagram with the dynamic model in Section 8.3.

We recollect our preliminary analysis, where the client expressed a desire for the staff to maintain a degree of control over the weaving process (Section 5.3.2). It might therefore be desirable for the control system to allow the loom operator to allocate a batch manually.

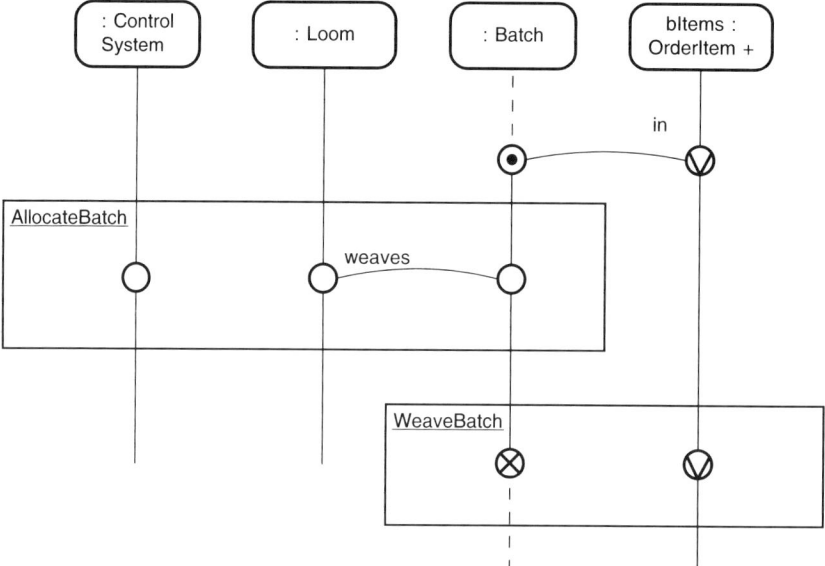

Figure 10.2 Weaving a Batch

The looms still expect a sequence of instructions, which, to produce a sensible ribbon, must be a combination of sequences of instructions corresponding to characters in a name, and instructions for the special separator characters. The old control box issued instructions for the various separators and special marks by virtue of its wiring; the new control system should ensure that these characters are woven at the appropriate point. The old control boxes also decided whether the following pick is to be ground or brocade; the new control system also has to do this.

So, in addition to the first task of allocating batches, the control system has to:

- translate the text of the batch items into appropriate sequences of instructions
- pass the instructions to the loom's electromechanical devices at the appropriate time

The translation of the batch item text entered by the operator is into sequences of instructions that correspond to the appropriate thread patterns for the characters in the text. This, as we noted in Chapter 8, is a central function of the control boxes. There are many acceptable ways in which this translation might be done, so we refrain from being too specific here—we just say that it has to be done. The only constraint is that the translation has to have occurred sometime in between a batch being created and a batch being woven.

Looking at our models so far, we can further extend the model given above to that shown in Figure 10.3. This allows several different scenarios concerning when

10.2 What the New World looks like

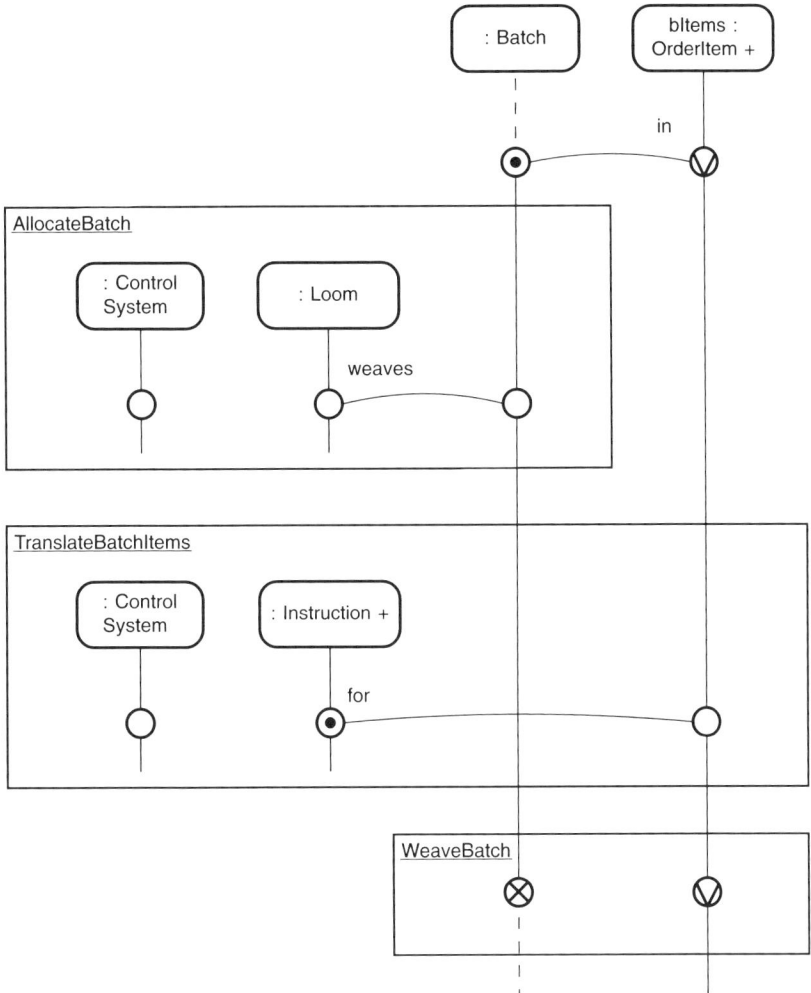

Figure 10.3 Translating a Batch Item

translation is done, since *AllocateBatch* and *TranslateBatchItems* are shown as independent behaviours (the vertical positioning is just for graphical convenience). If the control system is not busy, it could be translating items in advance of allocation. Alternatively, it might leave translation until the last minute (that is, just before weaving). These decisions have to be made in the light of any timing requirements. Decisions about how the translation is done also have to be made at some point, in the light of subsequent architectural decisions.

The required ability to expand into foreign markets is relevant here. A rich character set and the corresponding translations could be maintained by the sys-

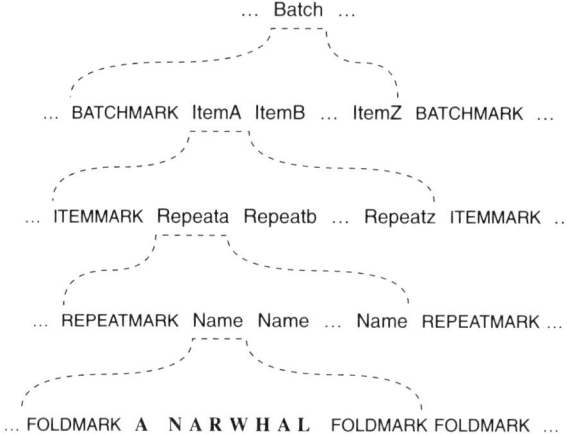

Figure 10.4 The structure of a batch of name tapes

tem. Alternatively (or in addition), the control system could allow the definition of new character sets and translations. The latter option would give considerable flexibility to the system.

As well as translating the text of each order item into loom instructions, the control system has to collate these instructions into weavable batches, using various separator marks. Figure 10.4 describes the structure of a batch. The 'terminal symbols' (characters and markings) are shown in upper case. Each of these terminal symbols requires a sequence of instructions involving ground and brocade picks. Each brocade pick must be woven on top of a ground pick (that is, without moving the ribbons on), and each text character has appropriate ground picks woven on either side to separate adjacent characters. The ability to change this 'name tape grammar' easily is fundamental to the requirement of being able to weave things other than name tapes in the future.

A point to note is that a loom weaves 72 ribbons simultaneously. It follows that the minimum quantity of a given name is 72, and so names are ordered in multiples of 72 or 'repeats'. The number of repeats specified by an order item needs to be checked by the translation process. If an item is for more than one repeat, the REPEATMARK and the duplicated name-text needs to be generated.

The third task of the control system is to communicate the weaving instructions to the looms, and interface with the electromechanical actuators and sensors, in accordance with the *OnePick* behaviour described in Section 8.2.1. In addition, the new control system has to deal with all the looms concurrently. Given that the control system is necessarily 'distributed' (centralised controller, many looms), there are additional reliability requirements concerning the transmission of instructions over a communications network. In particular, the central controller must be able to discover if any of the looms has become non-operational for any reason (mechanical breakdown, warp breakage, thread running out). In other words, the control system needs to be able to handle 'exceptional' behaviour as well as 'normal' behaviour.

10.2.2 The operator interface

We now need to consider the operator interface. Minimally, the control system needs an operator to enter batch details. These details are in accordance with the model of a batch and its batch items given in Section 10.2.1. This is where any similarity between the old and new control systems ends for the operator. Everything else is new territory, the operator being faced with a central control system controlling several looms and the ability to enter batches far ahead of their time of weaving.

One obvious consequence of the change is that the operator may want to remove a batch. Before, this was not an issue, since a batch would simply not have been entered. Since so much is unknown about how the operators might want to deal with the new opportunities and limitations afforded by such a control system, many of the operator interface requirements have to be developed during the development phase, by means of prototyping and other techniques. However, there are requirements that we can talk about now.

The control system needs some information about the looms to allow it to allocate batches fairly: which loom is using which colour, and perhaps when the colour was put in, so that the control system can estimate when it will run out.

However, the operator should be able to specify which batch goes on which machine, and when, in order to retain control over the weaving process, as desired by the client. This might be useful, for example, for orders that have to be woven urgently. An alternative might be to allow priorities to be given to each order by the operator, and then let the control system make decisions based on this information. This is something that can only be assessed during design, in conjunction with the loom operators, who can determine how easy to use and how beneficial such a feature might be. This is separate from the issue of whether such priorities are used by the control system itself in order to determine its scheduling; such a mechanism might be appropriate for the control system to use, but undesirable for the operator.

Another consequence of the client's desire for the operators to maintain control over the weaving process is that information should be given to the operators about what the control system is doing or planning to do. Only if the operators have this information can they act intelligently in controlling the weaving process. Again, it requires the involvement of the loom operators themselves to decide what information would be useful. Information that was previously available to them is no longer obvious (for example, the ability to say which loom a batch was to be woven on). The range of information that they may be interested in is now wider (for example, the number of batches allocated to particular looms).

Since the information required is unlikely to be difficult to produce or assimilate, we can say that, in principle, the control system should be able to display to the operator any information it holds regarding looms and batches. The operators may also be interested in the working of the control system itself. Previously, it was obvious to the trained ear whether or not a control box was about to break down; with an automated system, such ailments may not be so easy to detect.

If we are particularly worried about the information required by the operators (if, for example, the operator has to make critical decisions on the basis of this information), then this single aspect of the control system has to be subjected to a much lengthier analysis.

A similar requirement on the control system is presented by the need for statistics. For the chosen bundle of prescriptions, it was agreed that any information about looms and batches that could be obtained easily from the control system would be made available for monitoring purposes. Since this information has not been easily available before, there is a further question to be answered about what kind of information is useful, and how the information should be manipulated and presented in order to be of help. We can simply say here that basic status information on batches, orders and order items should be made available.

Given that the loom operators have enough information to allow them to know that they want to intervene and what needs done, they have to be able to intervene. Removing batches was mentioned above. Other operator tasks, taken from the old world purposive models, include requesting maintenance for a loom and changing the colour of a loom. Both of these require the ability to remove the loom from operation.

10.2.3 Reliability and performance

We now need to consider various issues regarding the reliability and performance of the proposed control system. These issues are only briefly discussed here; a more detailed consideration needs to take place as part of the system design activity.

What happens when things go wrong?

We need appropriate shut-down and recovery procedures for the looms and the control system itself. So, for example, the control system should keep a non-volatile record of any batch items that are being woven but whose weaving has not been completed. Subsequent re-allocation must take into account the need to remain within the 14 day turnaround time. The control system should also be able to detect, flag and possibly act on abnormal signals from its cam contacts (most obviously, no signal at all).

How much storage is needed?

This has two aspects: How many batches can have pending status at any one time? Should records be kept for statistical purposes, and if so, of what and for how long?

At what frequency do instructions need to get to the loom?

Our client gives us some possibly relevant figures: each loom, at its fastest, takes about 2 minutes to weave a 12 character name; a typical character is woven as a sequence of 16 instructions—8 ground and 8 brocade; 12 name characters plus special characters and spaces suggests that about 100 instructions have to be provided each minute. However, there are two factors here: the physical constraints of the weaving frame, and the typical make-up of names and characters. The former can be assumed fixed, but the latter factor may change, for example with the introduction of new character sets.

How fast does batch allocation need to be?

This depends on turnaround time: batch allocation should be done so as to allow the 14 day turnaround time on orders to be met. For each of these timing requirements, we ought to ask whether the most significant timing constraints are under the control of the control system. For example, the biggest constraint on achieving the 14 day turnaround time for each order is currently loom down-time, due to control box unreliability, but also due to routine colour changes. A control system that could allocate batches to looms intelligently might be faster than one that put its effort into supplying instructions at break-neck speed.

What capacities does the control system need to handle?

The client suggests that the target for a four-loom NIMWeC is about 1000 orders a day, and so this is what the control system should be able to handle. Given that there is space in the factory for six of the current style looms, then, in the interests of future expansion, the ability to handle 1500–2000 orders per day seems a sensible precaution.

Another 'design' issue that should be considered at this early stage is the overall architecture of the IT system. In particular, we need to consider the hardware/software mix, and the distribution of functionality across hardware.

There are a number of distinct tasks in which the control system is involved. Broadly speaking, these are

- batch entry and monitoring
- batch allocation
- translation
- weaving

These tasks involve different parts of the physical environment. Batch entry takes place centrally, while weaving taking place in at least four different areas of the factory. Batch allocation and translation need to take place at or between these locations. Some parts of the control system therefore have to be located centrally, while other parts are located at each loom.

The details of the architecture need to be decided within the design process. However, the fundamental nature of the control system is that it is distributed, and so communications are an issue, with regard to timing, capacity and reliability.

We would expect that only the last task, weaving, requires any special-purpose hardware development.

10.3 Onward into development?

In Chapter 9 we set out and evaluated various bundles of prescriptions. The client chose a course of action on the basis of the evaluation, and in this chapter we specified the proposed development in more detail. We now know a lot more about the nature of this development, and it is sensible to ask the client at this point whether the cost/benefit analysis still holds, and if this is really what is wanted.

Only one aspect of our cost/benefit analysis has changed. With the realisation that the loom operators should be involved in the development of the new control system, the cost of subsequent training may be less than we originally envisaged.

We present the New World specification to our client, and are told that this is still the right course of action. The client is impressed with the way the project has gone, and asks us to proceed apace with the development of the system. We, too, after this analysis project, are confident that the proposed changes and the new system will be a help to NIMWeC. Eager to get started, we set about designing a Development Process that will take us through design and implementation of the new control system.

Part III

Using the Modelling Languages

Chapter 11

Purpose and Behaviour

11.1 Why use models?

Chapters 4–10 have worked through an analysis case study in some detail. The main visible outputs from the analysis activity are in the form of *models*: information expressed using defined modelling languages. In this part we step back from the Basic Process and discuss the concepts of the two ORCA modelling languages, and how they can be used to aid analysis. Appendices B and C in Part VI give precise definitions of the modelling languages, and can be used for reference purposes.

Before dealing with the ORCA modelling concepts, it is worth emphasising the reasons for using models in analysis. As can be seen from the NIMWeC example, models provide a *vehicle for analysis*—a way of writing things down, an aid to understanding and a form in which information can be examined and manipulated. The modelling activity is valuable in itself, even if the resulting models are not required as deliverables from the analysis process. Attempting to model something is an excellent way of prompting questions and checking one's understanding.

Modelling languages should serve as a *tool for abstracting and simplifying complex situations*, helping an analyst express what is relevant to an analysis and what is not. Conversely, modelling languages should also have mechanisms for composing simpler components into more complex ones, so that an analyst can ask 'how do these things fit together?'.

A related function of modelling languages is to *collate and present disparate information*. For example, we often want to produce an overall description of a part of the world, in terms of its structure or the kinds of object involved. In some situations a global 'object model' or 'information model' is useful. However, the analyst should be careful that collating information into a single model produces some useful 'macro-reading'. For example, we might be able to see that complexity is relatively localised, or that some entities are central while others are peripheral, or that a certain structural pattern emerges.

Another role of models in analysis is to *support statements about requirements for change and rationale for change*. Models can supply precise and detailed meanings for the terms used and the ideas expressed. The models in themselves are not the primary output of the analysis process—the analysis process is not about constructing Grand Unified Models of Everything. Initially at least, modelling should be done in a piecemeal fashion, as and when required. Subsequently, it may be desirable to integrate model fragments for the purposes of checking or presentation.

The degree to which analysis models *provide material inputs to the design process* is a difficult issue. If design is pursuing object orientation, it is certainly feasible to 'carry over' classes and related information from the analysis models to design models. However, if this is taken to be the default assumption, it should not pass unquestioned. The classes that appear in an analysis model may be quite unsuitable as the basis of object oriented software. Conversely, it would be a mistake to allow design considerations to influence the content of analysis models.

A method such as ORCA cannot lay down hard-and-fast rules concerning the role of models and the use of modelling languages in analysis. What is provided is a set of conceptual and notational tools, together with ideas about their use, as discussed above. How exactly these tools are used needs to be determined by the analyst in the context of a particular analysis. In particular, how complexity is handled—by more modelling or by less—must be decided by the analyst. If an analysis is getting 'bogged down' in some aspect of modelling, then the usefulness of continuing that particular modelling activity should be questioned. It is always permissible to stop and ask 'how much more about X do we need to know?'.

This is especially the case with modelling dynamic behaviour. There is a significant danger of being overprecise and worrying about interaction mechanisms, rather than trying to capture the minimal constraints for valid behaviours. This is an aspect of modelling where striving to produce a 'complete, unified model' is often a mistake. In contrast, descriptions of static structure (classes and relationships) are generally more manageable, and it may be worth producing a 'complete object model'.

The conclusion of this section is that, for the purposes of analysis, modelling should be done judiciously, recognising that it should be the servant of the analysis process, not its master.

11.2 Purposive entities and Behavioural entities

The examples in Chapter 2 and Part II illustrate ORCA's dual modelling approach: purposive modelling and behavioural modelling. However, we have not, so far, said much in general about the different kinds of entity in the two sorts of model. The following example aims to clarify the distinction between purposive entities and behavioural entities. The example has the flavour of an analogy, and is intentionally nothing to do with IT development.

11.2 Purposive entities and Behavioural entities

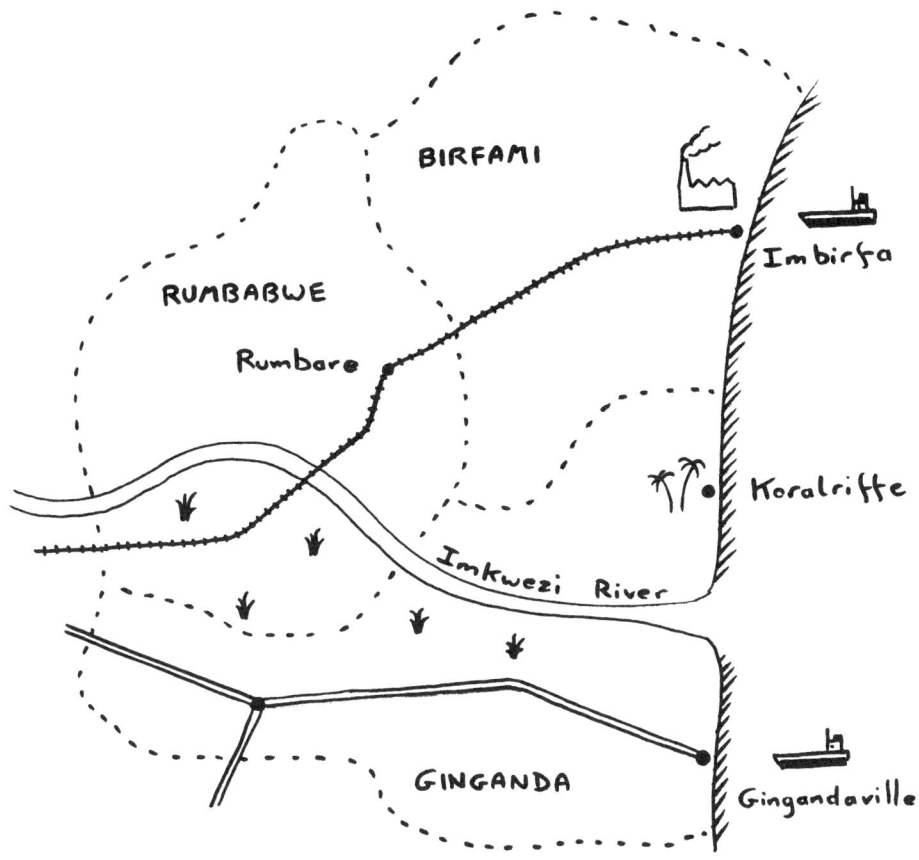

Figure 11.1 Rumbabwe, Birfami and Ginganda

Suppose that we are dealing with three fictional neighbouring countries: Rumbabwe, Birfami and Ginganda. The geography is sketched in Figure 11.1.

The south of Rumbabwe is dry, but very fertile. Alfalfa is grown there, thanks to extensive irrigation that takes water from the Imkwezi river. The alfalfa is Rumbabwe's main export, but since the country is land-locked, it needs to be transported by rail to Imbirfa, which has a deep-water port. Imbirfa also has large industrial complexes (some owned by multinational companies) that generate a variety of unpleasant waste products.

Ginganda shares the fertile southern region with Rumbabwe, and it, too, takes water from the Imkwezi river. The amount of water required for irrigation is increasing in both countries. The alfalfa is transported by road to the port at Gingandaville. The other main source of revenue for Ginganda is the group of beach resorts centred around Koralriffe; their slogan is *Come to Koralriffe for a*

Ginganda tonic. Since the prevailing ocean current flows from north to south, these resorts, famed for their crystal clear waters, are at risk from the industrial pollution of Imbirfa.

There are two quite distinct aspects here: the 'political geography' of countries and their interrelationships, and the 'physical geography' of cities, rivers, railways, and so on. The requirements and objectives of the three countries are imposed on the geography of the region. The regional structuring works effectively if these requirements and objectives are compatible, and if they can be satisfied by the physical behaviour of industries, agriculture, transport links, and so on.

We can thus treat the three countries as *purposive entities*, and look at the way that they are related. Rumbabwe relies on Birfami to operate the Rumbare–Imbirfa railway effectively, and Birfami may in return guarantee to do so. If Birfami does not co-operate then there is a problem. Similarly, Ginganda relies on Rumbabwe not to deplete the Imkwezi river to a level that jeopardises its own irrigation, and on Birfami to control discharges of industrial waste from Imbirfa. We might also want to regard the multinational companies as purposive entities, having relationships with Birfami in particular.

For the purposes of our regional analysis, the railways, rivers, industries, and so on, are *behavioural entities*. Rumbabwe relies on Birfami to provide a 'service' that concerns the effective running of the railway. Even if the two countries agree on the definition of the service to be provided, the behaviour of the railway may not actually realise the service. In order to determine the level of service, we could model the frequency and capacity of trains on the line, and relate this to the requirements of alfalfa production (quantities at different times of the year, how long it can be kept in storage).

A point worth noting is that the countries are not directly observable. We cannot go somewhere, look around, and see 'Rumbabwe'. On the other hand, we can directly observe railways, rivers, bales of alfalfa, and so on. Another point to note is that the behavioural entities are not partitioned by the purposive entities. The relationships between the countries are mediated by 'shared' behavioural entities—the river, the railway, the ocean currents. Behavioural entities that are entirely internal to a country are not directly relevant to the purposive relationships between countries.

Whether things are treated as purposive entities or behavioural entities depends on the nature of the analysis. What happens if we treat countries as behavioural entities? A behavioural model could contain three instances of class Country. This class can define properties that apply to countries *in general*, for example their capital city, population, gross domestic product, and the position of their borders. On the whole, these properties do not seem directly relevant to the purposive relationships between countries. A possible exception is the definition of a country's borders; if borders change, the purposive relationships may change (for example, if Rumbabwe ceases to be land-locked). This indicates that we should treat countries primarily as purposive entities.

11.2 Purposive entities and Behavioural entities

What do countries actually do? Firstly, they change the sort of properties mentioned above—changing capital city, increasing population. Secondly, they enter into agreements with each other, concerning trade, transport and environmental controls. In other words, they formalise purposive relationships. This is another indication that we should treat countries primarily as purposive entities.

Both purposive and behavioural entities can be structured—they can have component entities and containing entities. However, the nature of the structuring is different. For example, we might want to express the co-operation between Rumbabwe and Birfami at a lower level, as a co-operation between the Rumbabwean alfalfa producers and the Birfami State Railway Company. Rumbabwe and Birfami are treated as distinct entities with a *different* internal structure. In contrast, if we want to say more about countries as behavioural entities, then we model the class Country as a set of constituents. For example, we might choose an economic model, containing an Industrial Sector, an Agriculture Sector and a Transport Sector. All three instances of Country thus have the *same* internal structure. We might decide that the Rumbabwean alfalfa producers constitute its Agricultural Sector, and that the Birfami State Railway Company is that country's Transport Sector. However, we should be cautious about leaping to this conclusion.

On the other hand, some systems do seem to have levels where purposive and behavioural entities coincide (or could be made to coincide). In the example of NIMWeC, there could be Manufacturing and Orders & Invoicing divisions that have co-operation relationships with each other and with the company as a whole. They would be responsible for performing particular activities, and so provide the required services. The divisions of the company could thus be viewed as both purposive entities (roles) and behavioural entities (subsystems).

Why is there a (possible) coincidence of purposive and behavioural entities in NIMWeC, but not in our fictional geographical region? The obvious difference is that the organisation of NIMWeC can be designed, or at least evolved, to work effectively. In contrast, the political and physical geography of our fictional region is the product of complex historical and geographical processes.

Organisational and administrative structure is behavioural, not purposive. If the purposive entities map on to this structure, then all well and good. However, if there is a mismatch between these alternative structurings, then we can expect problems. If purposive entities have no behavioural equivalents there may be no way of establishing the necessary co-operations. For example, our fictional countries need mechanisms for drawing up and ratifying agreements to co-operate. This co-operation-establishing behaviour should be distinguished from the behaviour that realises the services involved (running trains, controlling pollution, and so on).

A system is going to work effectively only if purpose and behaviour are harmonised. An organisational division whose activities do not match its role in the overall system can hardly be effective. In general, we should expect to find a coincidence of purposive and behavioural entities (at some level of abstraction) in systems that have been explicitly designed. However, even where a system has been explicitly designed, historical and social processes may cause purpose and behaviour to diverge over time. The cumulative effects of such divergence are a likely subject for analysis

In conclusion, we can say that analysis involves the interplay between purposive and behavioural views of the world. This dual view helps an analyst to consider problems, underlying pathologies and candidate solutions.

Chapter 12

Behavioural modelling

12.1 Histories and Frameworks

A Beluga framework provides a behavioural model of a system (or part of a system) in terms of the possible histories of objects and their interactions. In practice, non-trivial systems can have an infinite number of possible histories.

If this sounds rather daunting, consider the NIMWeC system. What NIMWeC can potentially do is defined by its framework model. What NIMWeC actually does, the particular history that it exhibits over a given period of observation, depends on when customers generate orders, on the quantities, styles and details of the name tapes ordered, on the operation of the weaving machines, and so on. In principle, we could record what happens in terms of the creation, association and interaction of objects in the history. However, NIMWeC could exhibit many different actual histories while still being the same system.

A behavioural model thus describes what is behaviourally *essential* to a system. The model characterises the possible histories in terms of *patterns* of object creation, association and interaction. Since behaviour happens over time, these patterns have a time dimension, as well as an 'object space' dimension (what makes different objects different).

We can illustrate this using a simplistic Lending Library, in which borrowers can take out and return books (without limit). When a book is on loan to a borrower, then it is *associated* with that borrower—physically in their possession, and recorded by the library as being so. When a book is back in the library, it is not associated with any borrower. The obvious physical constraints mean that a book can be on loan to no more than one borrower at a time.

Suppose that we observe the behaviour of a particular (rather small) library for some period. We could record the history of the system as shown in Figure 12.1. This scenario shows a succession of events, with time progressing down the page. Each object is shown by a vertical *timeline*. Events are either *associate* events (borrower takes out book), or *dissociate* events (borrower returns book).

122 Chapter 12 Behavioural modelling

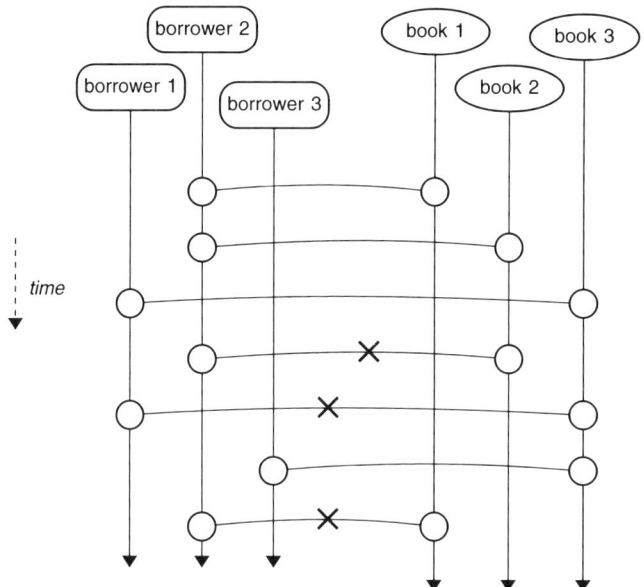

Figure 12.1 Behaviour of a small library

Associate events are shown by a horizontal line connecting a book timeline with a borrower timeline; dissociate events are shown similarly, but with a cross on the line (indicating that the association is broken).

This Lending Library can exhibit many different histories, depending on which books and borrowers are involved, when borrowers decide to take out and return books, and which books are selected by borrowers. A behavioural model needs to provide a pattern that fits all allowable histories, while excluding those histories that are not allowable.

The key behavioural element is the *Loan*, consisting of a borrower taking out a book, and subsequently returning it. Any history exhibited by the Lending Library consists of a set of *Loans*. Where different books are involved, loans can overlap in time, but a book can be involved in no more than one loan at a time. (Figure 12.2).

This behavioural pattern can be expressed as in Figure 12.3. This framework says that a *Loan* involves a *Book* and a *Borrower* in two events, *takeOut* and *return*. A *Library* behaviour consists of zero or more ('*') *Loans*, which may be taking place in parallel. However, a book's change of state means that a book cannot be involved in more than one loan at a time, since it can be taken out only if it is *in*.

The round-ended boxes containing : *Book* and : *Borrower* represent the *stereotypical* book and borrower objects involved in a *Loan*. Different loans involve

12.1 Histories and Frameworks

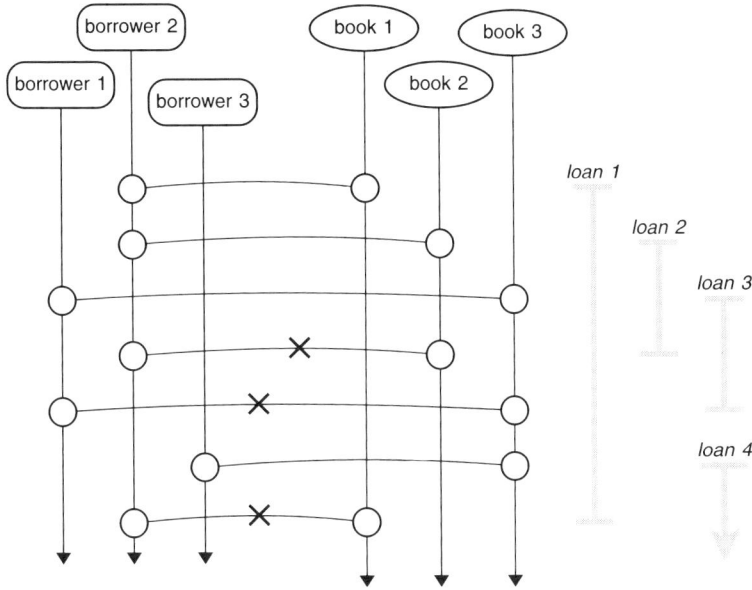

Figure 12.2 Loan timelines for a small library

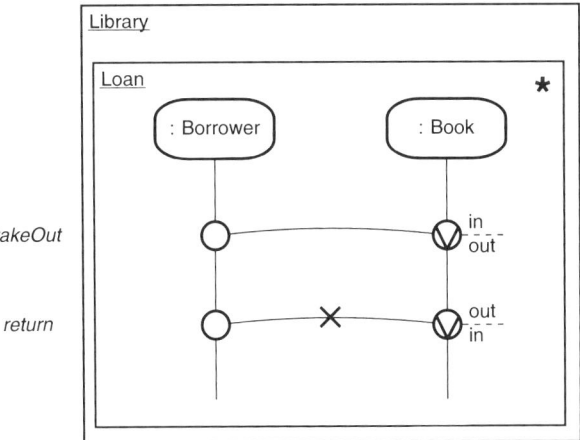

Figure 12.3 Loan behaviour

different books and borrowers. Within a single *Loan*, the book and borrower participate in the associate and dissociate events. These events are shown by horizontal links between participation 'blobs' on the object timelines. The positioning of the dissociate event below the associate event on the timelines indicates that the former event has a time *dependency* on the latter—within a *Loan*, a *takeOut* always happens before a *return*.

12.2 Object characterisation

12.2.1 Classes and instances

In modelling our Lending Library we characterise the objects as either Books or Borrowers. In the terms of object orientation, we have a *class* Book, and a class Borrower, with many *instances* of these classes. If we want to name an instance in a particular context, we write *someParticularBook* : *Book* or if we do not need to refer to the instance by name, we can write : *Book* as we did in our model of the Lending Library. This can be read as 'a book'.

Instances of class *Book* and of class *Borrower* are distinct—an object is either a book or a borrower. Objects never change classes, although in some circumstances they can be treated as instances of different classes (this is discussed later).

12.2.2 Object states

Properties of an object that can change are regarded as the state of that object. For example, a book can change from being *in* the library to being *out*, and back again; at any given time, some books are *in* and some are *out*.

Notice that the named statuses *in* and *out* are just a convenient way of saying whether a book is currently associated with a borrower, or not. The statement 'a book can be taken out only if it is in' is equivalent to the statement 'a book can be taken out only if it is not currently associated with a borrower'. We can also regard the *in/out* status of a book as 'remembering' whether the most recent event involving the book was a *takeOut* or a *return*.

The state of an object can reflect its initial state and its entire history (that is, all the events in which it has been involved). For example, a book could have an attribute (which takes NATural numbers as its values)

> *numberOfTimesBorrowed* : NAT

This is used to reflect the number of *takeOut* events in which the book has been involved; its value increases by one every time the book is taken out.

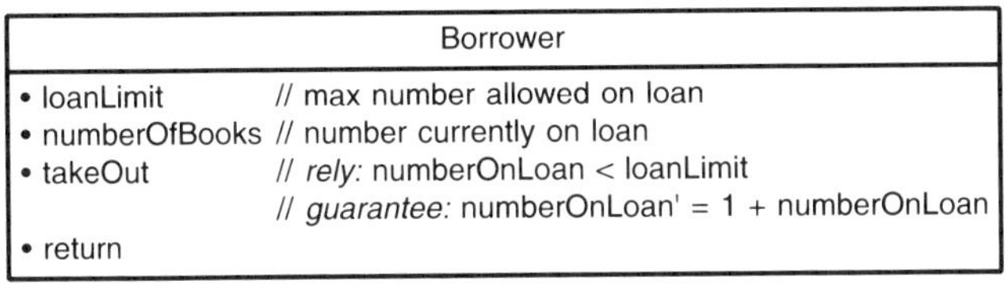

12.2 Object characterisation

```
┌─────────────────────────────────────────────────────────────┐
│                         Borrower                            │
├─────────────────────────────────────────────────────────────┤
│ • loanLimit        // max number allowed on loan            │
│ • numberOfBooks    // number currently on loan              │
│ • takeOut          // rely: numberOnLoan < loanLimit        │
│                    // guarantee: numberOnLoan' = 1 + numberOnLoan │
│ • return                                                    │
└─────────────────────────────────────────────────────────────┘
```

Figure 12.4 Features of Borrower

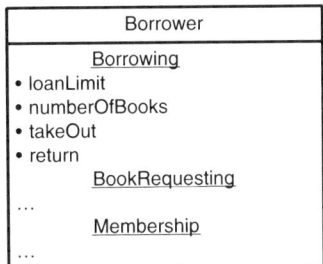

Figure 12.5 Facets of Borrower

If we want to limit the number of books that a borrower can have at any one time, *Borrower* would need the attributes

$numberOnLoan : \text{NAT}$

$loanLimit : \text{NAT}$

numberOnLoan increases by one for every *takeOut* event involving the borrower, and decreases by one for every *return* event. A *takeOut* event involving a borrower would be allowed only if the borrower's loan limit had not been reached:

takeOut
  // rely : $numberOnLoan < loanLimit$
  // guarantee : $numberOnLoan' = 1 + numberOnLoan$

$numberOnLoan'$ refers to the value of the attribute after the operation has been performed.

12.2.3 Features and facets

In our model, the class *Borrower* would now have several features—attributes, and operations that modify the values of attributes (Figure 12.4).

Borrowers might have features that concern different aspects of behaviour—borrowing, requesting new books, membership of the library, and so on. We can

126   Chapter 12 Behavioural modelling

Figure 12.6 Association of Book with Borrower

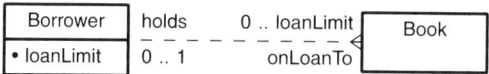

Figure 12.7 Association of Book with Borrower, with cardinalities

Figure 12.8 Borrower, with 'books-held' attribute

therefore group features into *facets* (Figure 12.5). In fact, we might well identify the facets before the features.

A useful heuristic is that the names of facets can often be gerunds, that is, a noun formed by adding '-ing' to a verb. In the above example, we have 'borrowing' and (book) 'request-ing'. 'Membership' is to do with the relationship between a borrower and the library—'establishing', 'maintaining' and 'terminating' it, presumably.

12.2.4 Static structure

Having characterised the objects in the Lending Library as either books or borrowers, we might want to see a *static view* of the system, in terms of the classes of object and the ways in which instances of these classes can be associated. In this case, a book may be associated with one borrower; a borrower may be associated with one or more books. This is drawn as in Figure 12.6. The dotted line indicates that the association is optional at both ends; the crows-foot indicates multiplicity at the Book end.

We could name the association at either end, and we could also be more precise about the multiplicity: (Figure 12.7). Notice how the diagram can be read from left to right as 'borrower holds zero to *loanLimit* books', and from right to left as 'book *onLoanTo* zero to one borrowers'.

Do we need this association? Couldn't Borrowers just have a *booksHeld* attribute (Figure 12.8)? *booksHeld* is a set of zero or more ('*') books. However, this is just one form that the association between borrower and book might take. Equally well, a book might record the name of its current borrower, if any (Figure 12.9).

12.2 Object characterisation

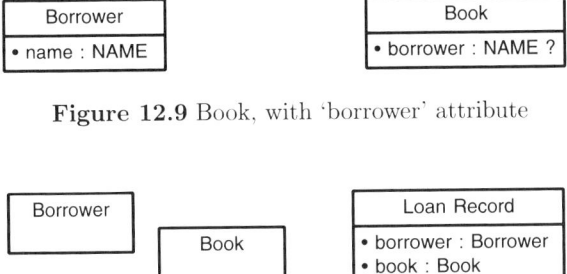

Figure 12.9 Book, with 'borrower' attribute

Figure 12.10 A Loan Record

Here, borrowers have a *name* attribute (of some basic type NAME), and books have a *borrower* attribute that may ('?') have a value of the same type.

Yet another option is to have an object of some third class—*LoanRecord*, say—capture the association by having both *book* and *borrower* attributes (Figure 12.10). A loan record would be created every time a *takeOut* event happens, and would be deleted when the corresponding *return* event happens.

We might have any combination of one or more of these three arrangements. In a real library, we would probably have the first and third cases—borrowers would hold books, loan records would be kept by the library, but books would not record borrower identities. Using an association is thus more abstract than using attributes. If using an association captures what is 'essential', then the arrangement that realises the association is an aspect of the world that could be changed.

We can take a static view of *interactions*, as well as associations. This is discussed in a later section.

12.2.5 Generalisation and specialisation

Objects are not always instances of completely unrelated classes such as *Book* or *Borrower*. Sometimes we want to characterise objects as instances of a class that specialises another, more general, class. Suppose that our Lending Library has members who may read the books in the Reading Room, and some special members, called borrowers, who may also borrow books (Figure 12.11). In this case, the library has instances of both the general class and the specialised class.

We might want to relate two classes by a common generalisation that itself has no instances. For example, the library has items, which are either books or journals (Figure 12.12). Because there are no instances of *Item* that are not either books or journals, the class Item is referred to as an *abstract class*.

The properties that books and journals have in common (concerning acquisition and cataloguing, say) can be expressed as features of the *Item* class (Figure 12.13).

128   Chapter 12   Behavioural modelling

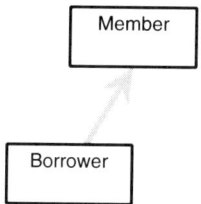

Figure 12.11 Borrower specialises Member

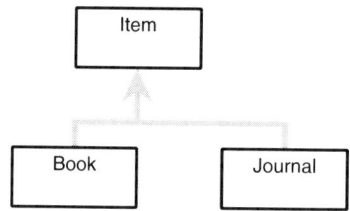

Figure 12.12 Book, Journal specialise Item

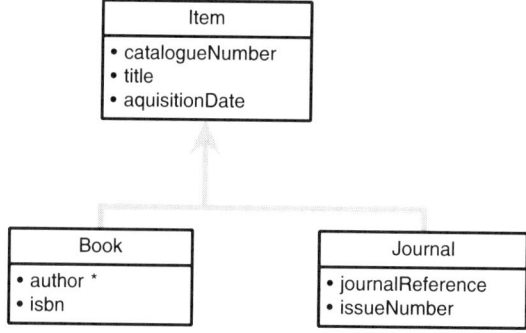

Figure 12.13 Book, Journal inherit features of Item

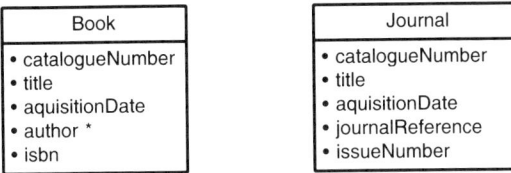

Figure 12.14 Flattened definition of Book and Journal

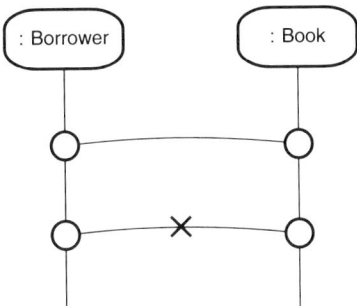

Figure 12.15 Association and dissociation during a Loan

Instances of *Book* and *Journal* are said to *inherit* the features of *Item*.

We could instead have defined *Book* and *Journal* as in Figure 12.14. However, our behavioural model cannot now talk about events that involve general items, either books or journals, so using the common abstract class gives us greater expressiveness.

12.3 Dynamic behaviour

12.3.1 Associate and dissociate events

We have already encountered associate and dissociate events in Section 12.1, in our model of a Loan (Figure 12.15). Here, an *associate event* is followed by a *dissociate event*. The former creates a link between an instance of *Borrower* and an instance of *Book*; the latter removes the link. This link is an instance of the *association* between book and borrower given by the static view (Figure 12.7).

If there were more than one association defined between books and borrowers (for example, 'book *reservedFor* borrower'), the kind of link created by an associate event or removed by a dissociate event could be stated explicitly (Figure 12.16).

The cardinality constraints defined statically for the association must be maintained by the composition of all loans. Since instances of the 'borrower *holds* book'

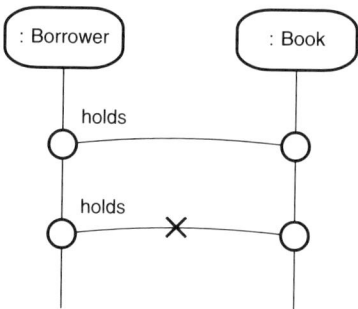

Figure 12.16 Named Association

association are created and destroyed by loans, the association is clearly optional at both ends (a book may be *onLoanTo* a borrower; a borrower may *hold* a book). The constraint that a book may be on loan to no more than one borrower can be represented either as a change of status (a mutation) from *in* to *out* and back again

[*borrower—book* : *hold*,
mut{*in* ⇒ *out*}*book*]
...
[*borrower*-x-*book* : *hold*,
mut{*out* ⇒ *in*}*book*]

or as a reliance of the *takeOut* event (that for the book concerned, it must not currently be held by another borrower)

"book not held"?(*borrower—book* : *hold*)
...
borrower-x-*book* : *hold*

12.3.2 Interactions

We have seen that dynamic behaviour consists of events, and that these can be *associate* or *dissociate* events. Since classes can define operations on objects, we also want to talk about events in which operations are invoked, and attribute values are changed as a result. These *interact* events can be described informally as 'object *a* does operation *b* to object *c*', where *b* is defined by the class of *c*. Sometimes we want to less precise and say 'something does *b* to object *c*' or 'object *a* interacts with object *c*'.

For example, suppose that *Borrower* has the feature *setLoanLimit*, which is an operation to set the *loanLimit* attribute to some value. We can say that this operation may be invoked by a librarian (Figure 12.17). If we had identified the

12.3 Dynamic behaviour

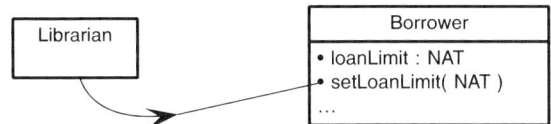

Figure 12.17 Librarian invoking *setLoanLimit*, static view

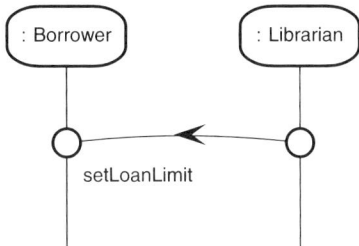

Figure 12.18 Librarian invoking *setLoanLimit*, dynamic view

facets of borrower before its features, we could have shown the interaction going to the *borrowing* facet rather than to a specific feature.

If we want to talk about when occurrences of this interaction happen, we can show the interact event by connecting object timelines, in a similar manner to that for associate and dissociate events (Figure 12.18).

Notice that we have two views of the 'librarian sets borrower's loan limit' behaviour. We can express the interaction as a static relationship between classes: 'librarians (sometimes) set borrowers' loan limits' (Figure 12.17). We can also talk about occurrences of this behaviour as events involving an instance of *Librarian* and an instance of *Borrower* (Figure 12.18). *setLoanLimit* events can take place before, after or in parallel with events of other kinds.

Rather than invoking operations, interact events may cause the target object to change *status*. Suppose that librarians can prevent borrowers from taking out books (for some misdemeanour, say), by making them *suspended*; suspended borrowers can subsequently be re-instated as *active* (Figure 12.19).

Notice the distinction between this pair of interact events and the associate–dissociate pair of events that form a *Loan*. In the case of a *Loan*, the association changes the state of both the borrower and the book. The associate event does not say how the association comes about—it might come about by a librarian creating a loan record, rather than through direct operation on either book or borrower. In contrast, the interact event that suspends a borrower does not affect the state of the librarian, who is merely the agent for the state change of the borrower.

Interact events may have *parameters*. This allows us to describe events of the form 'object *a* does operation *b* to object *c* with object *d*', or 'object *a* passes

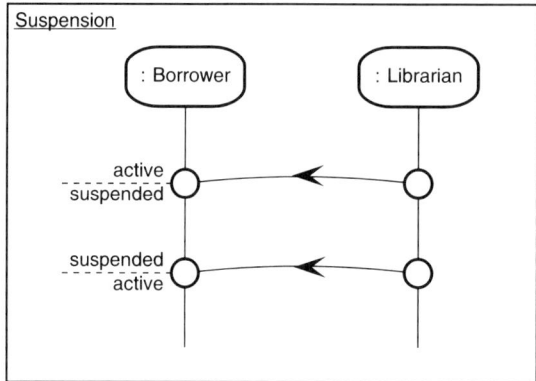

Figure 12.19 Suspending a Borrower

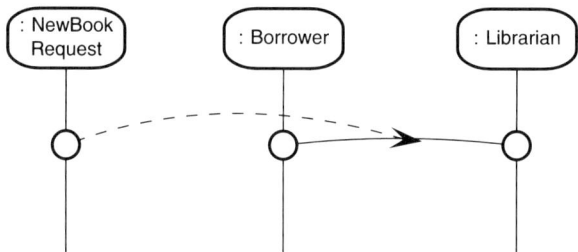

Figure 12.20 New Book request

object d to object c'. For example, a borrower might give a new book request to a librarian (Figure 12.20). After such an event, the librarian 'knows about' the new book request, and can perform operations on it, or perform behaviours that depend on its attributes.

12.3.3 Initiation and termination

While some objects may appear as permanent constituents within an episode of a history, other objects may be created dynamically during the episode. An object may be *permanent* or *transient* relative to the lifetime of other objects. For example, borrowers may be permanent relative to loan records but transient relative to the library as a whole.

In describing dynamics, we may therefore want to talk about object *initiation* and *termination*. We use these terms in preference to 'creation' and 'destruction', since we are interested in behavioural history, not in physical existence. The first thing that happens to a borrower in the Lending Library is their enrolment; our

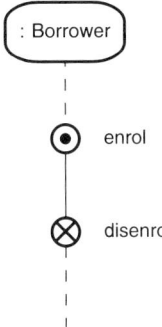

Figure 12.21 Initiation and termination of a Borrower

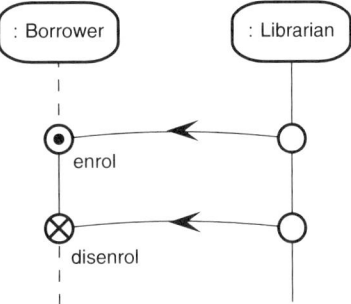

Figure 12.22 Enrolling and disenrolling a Borrower

model states that an *enrol* event initiates a borrower. The physical beings that our borrower objects represent are certainly 'created' prior to enrolment (unless the high demand for library membership requires education-conscious parents to enrol their offspring at birth). Similarly, disenrolment of borrowers is not usually accompanied by summary execution. Initiation and termination events are shown in Figure 12.21; the dotted portions of the timeline help to indicate the extent of the object's lifetime.

Initiation and termination may be an aspect of interact events. For example, it is the librarians who enrol and disenrol borrowers (Figure 12.22).

Sometimes, we may be interested in objects that are involved in only one event in their lifetimes. A typical example is that of a 'signal' object that interacts just once with some receiver object; the type or state of the signal determines what the receiver object does in response (Figure 12.23). Note that the symbol for a transient object is just a combination of initiation and termination symbols.

134   Chapter 12   Behavioural modelling

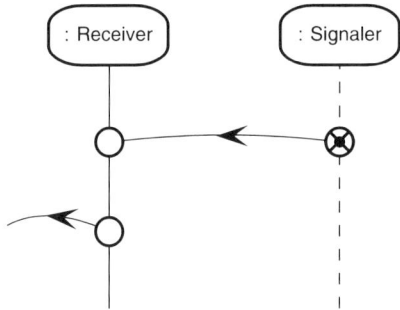

Figure 12.23 A transient signal event

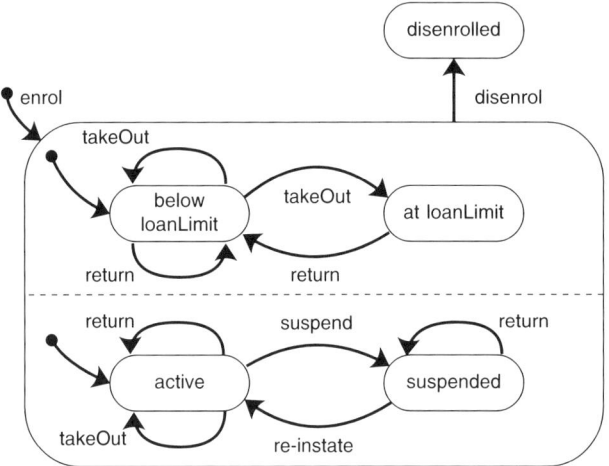

Figure 12.24 Harel statechart for Borrower

12.3.4 Object histories

In our library example, events involving borrowers typically result in changes to their status; this in turn constrains subsequent behaviour. For example, suspension prevents a borrower from taking out books; if a borrower takes out a book, and so reaches their loan-limit, further take-outs are prevented until they have returned a book. In this situation, it can be useful to focus on individual objects and describe their *life history* in terms of possible sequences of events.

Using ORCA does not restrict you to using only the Grampus and Beluga notations. Other techniques are also useful, and may be used. One approach that is useful for showing the state changes that occur for a single object is to use Harel statecharts [Harel 1987], a development of traditional state transition models. The

Harel statechart for *Borrower* is shown in Figure 12.24. Such a statechart notation is not part of ORCA's modelling languages; it has its own diagrammatic conventions.

This statechart shows two top level states for borrower—*enrolled* and *disenrolled*. A borrower is enrolled and disenrolled only once (if the same person re-enrols, they count as a different borrower). Inside the *enrolled* state, there are two independent components—whether or not the borrower is currently at their loan limit, and their active/suspended status.

Arrows between states represent the atomic events that a borrower may undergo. The first event that happens to a borrower is *enrol*; this results in the compound state *belowLoanLimit+active*. In this state, *takeOut* and *return* events are allowed, and do not change the state until such time as a borrower reaches their loan limit. When this happens, a *return* event must happen before another *takeOut*. If *suspend* happens to a borrower, *takeOut* events (but not *return* events) are prevented until *re-instate* happens. Suspension does not affect the state with regard to the loan limit.

This kind of model complements the view of dynamics taken by Beluga: behaviour of an object is clustered by its class, rather than objects being related by types of event. It is worth noting that the statechart does not tell us what *takeOut* and *return* events do, namely, associate and dissociate an instance of borrower and an instance of book (although these event names could also appear in a statechart for *Book*). Neither does the statechart say that *return* must return a book that has previously been taken out and not previously returned. Also, the statechart is more sequential than the Beluga model, since it does not allow a borrower to take out or return multiple books at the same time—*takeOut* and *return* events must be regarded as happening in sequence.

12.3.5 Constituent sets

All the events illustrated so far in this chapter have involved single instances of classes—a borrower, a book, a librarian, and so on. In general, however, events involve *constituent sets* which can have multiple elements. For example, we might have associate events involving sets of objects (Figure 12.25). Here, a set of one or more books becomes associated with a single subject category. That is, each book in the set becomes associated with the subject category. An associate event involving two sets results in all possible associations between elements of one set and elements of the other.

Sets of elements can also be targets (but not sources) for interact events. For example, we might describe an event in which a librarian suspends all borrowers who have failed to pay their subscriptions.

Constituent sets do not have to be disjoint. We might define one set in terms of one or more other sets. For example, the Library might send out notices to all its borrowers reminding them to renew their membership; however, only some of

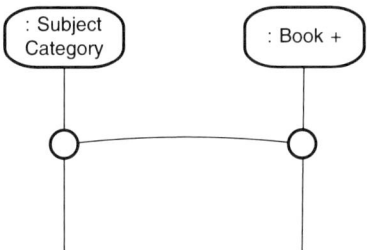

Figure 12.25 A set of Books in a Subject Category

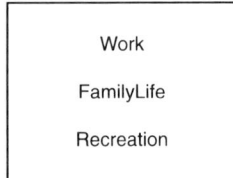

Figure 12.26 Composing processes

these borrowers would respond with a new subscription, the others would be automatically disenrolled. We would thus need to talk about three sets: *allBorrowers*, *renewingBorrowers* and *retiringBorrowers*. The second and third sets would partition the first set. This constraint can be expressed as a simple relationship between the sets:

$$allBorowers = renewingBorrowers \cup retiringBorrowers$$

12.3.6 Composing events

In talking about dynamic behaviour, we typically need to model complex behaviours in terms of simpler ones. For example, in our simple lending library model (Figure 12.3), a pair of *takeOut* and *return* events form a *Loan* episode; multiple *Loans* combine to form the behaviour of the library as a whole. All occurrences of *Loan* taken together could be regarded as a single long-running *process*. There might well be other processes, for example dealing with the enrolment and suspension of borrowers. The behaviour of the library would then be a composition of processes dealing with book loaning, library membership, and so on.

To express this structuring, we need a construct that allows us to *compose* frameworks to form larger frameworks.

In the real world, different things can be happening at the same time—different behaviours can be going on *concurrently*. For example, I might model my lifestyle

12.3 Dynamic behaviour   137

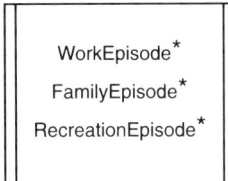

Figure 12.27 Composing multi-event processes

```
┌┃─────────────────────┃┐
 ┃ WorkEpisode*        ┃
 ┃ FamilyEpisode*      ┃
 ┃ RecreationEpisode*  ┃
└┃─────────────────────┃┘
```

Figure 12.28 Sequentially composing multi-event processes

in terms of three processes (Figure 12.26). The processes are named for conciseness, rather than being given explicitly (this would be done elsewhere in the model). The rectangular *composition* box groups the three processes into a larger behaviour.

Each process consists of multiple episodes (Figure 12.27), where the asterisks indicate multiplicity, as usual. Although I regard the processes as concurrent, the episodes that make up the processes may need to be interleaved, because I am unable, in practice, to do more than one thing at a time. For example, my typical day might consist of the following episodes:

- go for early morning jog (Recreation)
- have breakfast *en famille* (Family Life)
- go to work (Work)
- attend business meeting (Work)
- play squash at lunch-time (Recreation)
- write ORCA book (Work)
- go home (Work)
- cook dinner (Family Life)
- read children bedtime story (Family Life)
- play piano (Recreation)

...and so to bed. In this case, we could make it clear that the episodes happen sequentially by using a *sequential composition* (Figure 12.28). The double sides to the composition box indicate that the component episodes are 'squeezed' into a sequential stream.

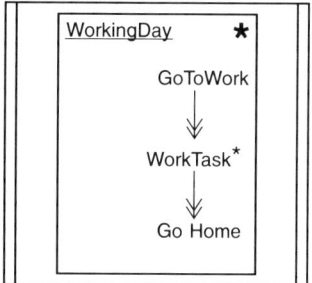

Figure 12.29 Sequential composition with temporal dependencies

This model does not constrain the order in which the various episodes happen. So, for example, it allows me to go to work repeatedly without coming home. The model therefore needs to express the *temporal dependencies* between events. I can say that my *Work* process is a succession of *WorkingDay* episodes. Each working day, I go to work, then do a number of work tasks, then go home (Figure 12.29). This model says that a *WorkTask* requires *GoToWork* to have completed, and that *GoHome* requires the *WorkTask*s to have completed (I can't do half a task, by definition). The model does not say what determines how many *WorkTask*s get done, just that they happen between *GoToWork* and *GoHome*.

The episodes within *WorkingDay* form a simple sequence, but this is not always the case. In general, we can have a network of dependencies in which independent events can happen concurrently. For example, suppose that I am cooking spaghetti bolognese. I might have the episodes and dependencies as shown in Figure 12.30. Episodes that are not linked by dependencies can potentially be done at the same time. For example, my sous-chef could be cooking the spaghetti sauce while I am sampling the Chianti. Similar kinds of dependency network are found in various planning techniques (PERT charts, for example).

In some cases we may want to compose events, for conciseness as much as anything. For example, we often want to talk about a composition of object initiation and association (Figure 12.31). This describes an event in which a borrower initiates a request for a book, where the request becomes associated with both the borrower and the book.

Sometimes it is useful to talk about a bit of behaviour as involving two or more objects (generally, constituent sets), but without going into details of associations or interactions. For example, Figure 12.32 shows a vaguer version of the previous model. Or we could extend the description by adding involvement of the librarian (Figure 12.33).

Ultimately, we might want to be more precise. A more concrete scenario would be as follows. First the borrower asks for, and receives, a 'blank' request from the librarian; then the borrower fills in *borrower* and *book* fields (that is, makes the associations); then the borrower submits the request to the librarian. However,

12.3 Dynamic behaviour   139

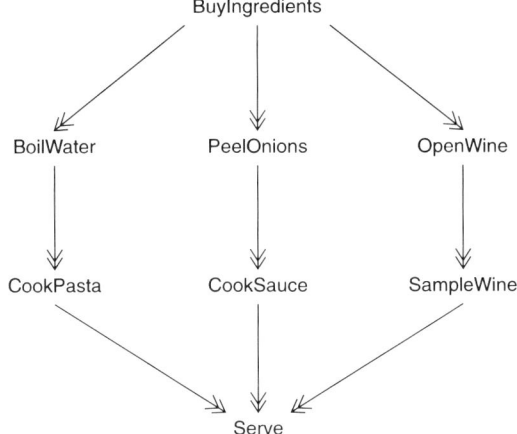

Figure 12.30 A network of temporal dependencies

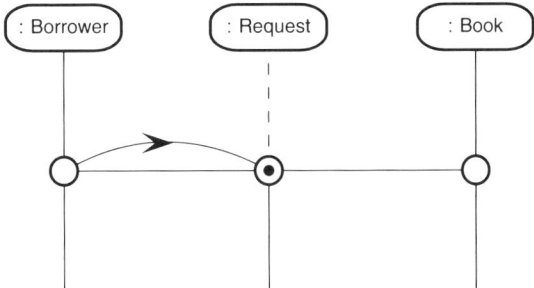

Figure 12.31 Composing initiate and associate events

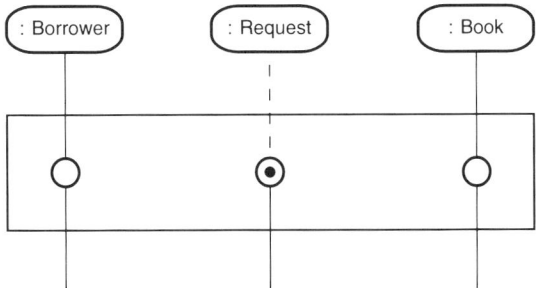

Figure 12.32 Composing participations

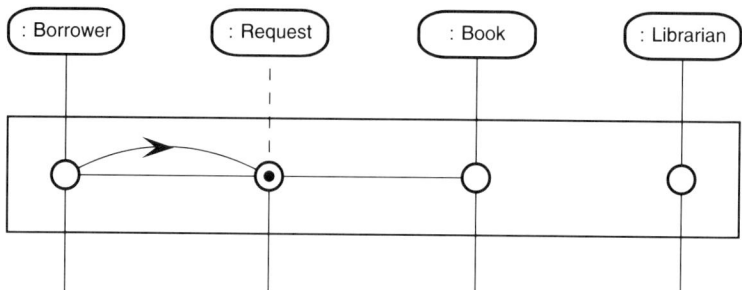

Figure 12.33 Composing events and participations

this level of detail may not be essential—it may be something that we are going to change and so do not want to model.

12.4 Levels of abstraction

In our Lending Library example, the objects involved—primarily borrowers and books—are fairly obviously 'atomic'. To model the essential behaviour of the lending library, we do not need to know about the anatomy of borrowers or the physical binding of books. However, there are some kinds of real-world entity that we may want to treat both as unitary objects and as compound systems.

Typically, we find this with machines and with human organisations. With machines, we may want a description of behaviour in terms of the machine as a whole: 'the operator starts the machine', and so on. On the other hand, we may want to talk about interactions of external entities with parts of the machine, and internal interactions between parts of the machine: 'the operator sets the gear lever and presses the start button', 'when the start button is pressed the ignition light comes on and the starter motor is turned on'.

A similar situation can arise when we are dealing with human organisations. Behaviour may be described in terms of entities at different levels of organisational structure: Company, Department, Team, Individual.

What we have are models at different *levels of abstraction*. A model that describes a machine as an object is at a higher level of abstraction than a model that treats the machine as a collection of interacting components. In the second model, the components are treated as objects, but they in turn can be modelled as collections of interacting subcomponents, and so on. In principle, there is no lowest level of abstraction, but in practice there is an 'atomic' level below which it is not sensible to decompose: human beings, nut and bolts, names and numbers.

One of the tasks of analysis is to find the best level of abstraction at which to describe behaviour. Too low a level and one cannot see the wood for the trees; too high a level, and one cannot see the wood *or* the trees. For example, in NIMWeC, it would not have been sufficient to treat *Loom* as an object, because we wanted to identify the parts that needed to be changed, and the parts that had to stay the same. On the other hand, describing how the electromechanical components of the jacquard or the patch panel worked would have been irrelevant to the proposed changes (the jacquard is not changing, the patch panel is being replaced).

12.4.1 An example of abstraction

We can illustrate these ideas using the Petrol Station from Part I. Suppose that we are interested in the interaction between customers and pumps, perhaps because we want to specify a new kind of pump. As a first approximation, we might describe the behaviour simply in terms of *Pump*s and *Customer*s. A Customer's transaction with a pump has the following form:

- The customer may change the setting of the 'grade' selector (for example, from '4-star' to 'unleaded'); this might happen more than once, if the customer changes their mind about the setting.
- The customer starts dispensing petrol.
- The customer may stop and re-start dispensing petrol, any number of times.
- The customer stops dispensing petrol.

Our model of this would be as in Figure 12.34. We really need to know a bit more about these interactions. For example, it is not clear how the *volumeDispensed* and *cost* attributes are changed. Also, how does the pump know when the final *stopDispensing* takes place?

To find out more, we can model a pump object as a collection of interacting parts:

- A *GradeSwitch* alters the connections of the *Line* that takes petrol from the storage tanks; the *GradeSwitch* sets an indicator light on the pump *Display* to show the grade selected.
- Withdrawing the 'nozzle' from the *Holster* starts the pump *Motor*; replacing the nozzle stops the *Motor*.
- Squeezing the *Trigger* on the nozzle engages the *Clutch* on the pump transmission; releasing the *Trigger* disengages the *Clutch*. The *Trigger* is accessible only when the nozzle has been removed from the *Holster*.
- If the *Clutch* is engaged and the *Motor* is running, petrol flows through the *Line*; a metering device in the *Line* sends pulses to the *Display* for every 0.1 litre; the *Display* calculates and displays the cumulative *volumeDispensed* and *cost*.

A static model of the *Pump* framework is shown in Figure 12.35. The 'missing link' between *Line*, *Motor* and *Clutch* is the physical flow of petrol. The dynamics might be as shown in Figure 12.36.

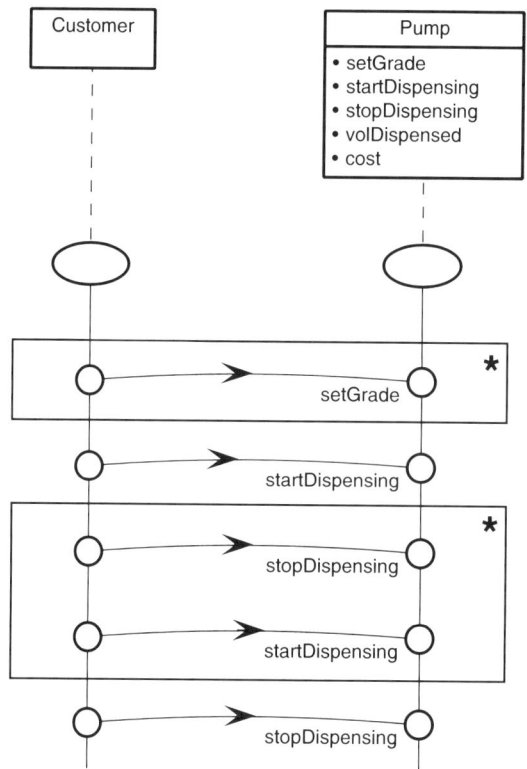

Figure 12.34 A Customer's transaction with a Pump

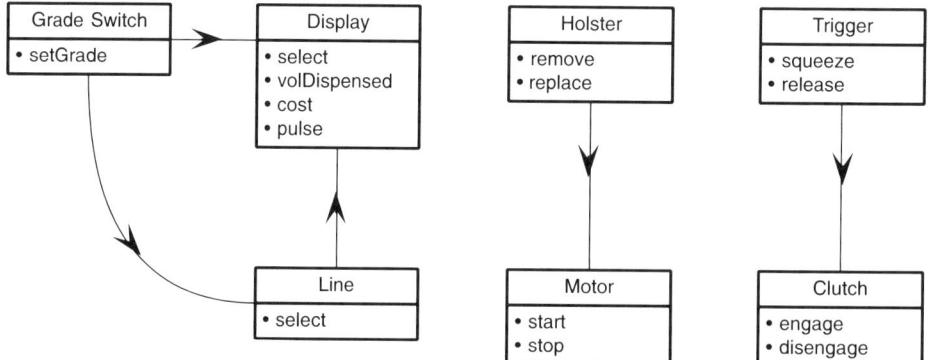

Figure 12.35 A static model of the Pump framework

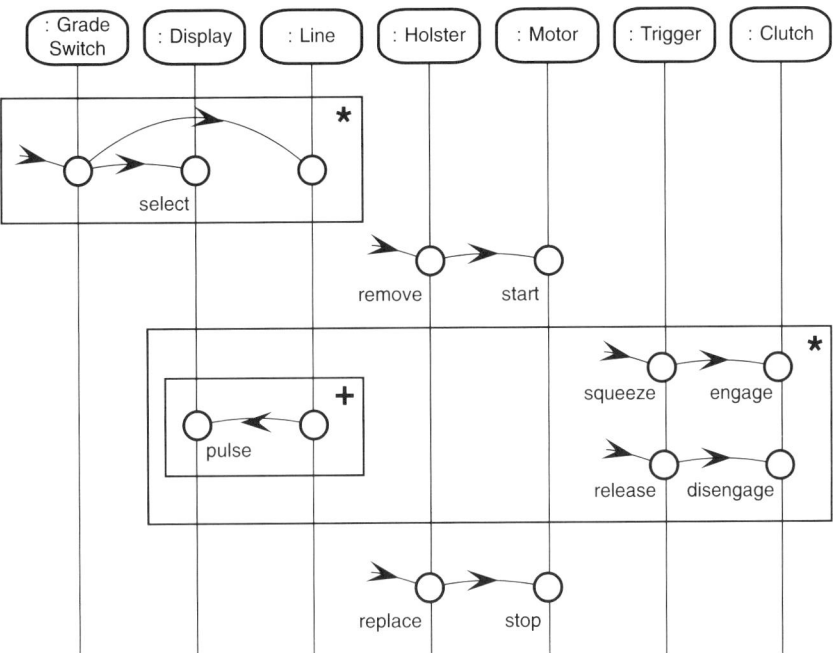

Figure 12.36 A dynamic model of the Pump

We can now see how the operations on the *Pump* class map on to operations on its component parts. At the higher level of abstraction, the customer invokes the *setGrade* operation on the pump object. At the lower level, the customer interacts with the grade switch, which in turn interacts with the display and the line. The initial *startDispensing* operation on the pump object becomes a sequence of two operations: *remove* on holster, and *squeeze* on trigger. Subsequent *startDispensing*s are just a *squeeze*. Similarly, the final *stopDispensing* operation on the pump object becomes a sequence of two operations: *release* on trigger, and *replace* on holster. Previous *stopDispensing*s are just a *release*. The *volumeDispensed* and *cost* attributes of the *Pump* class become attributes of *Display*; these are incremented by the *pulse* operation on *Display*, invoked by *Line*.

We thus have two models of a petrol pump: a Pump *class*, with attributes and operations defined for it, and a Pump *framework*, with constituent objects and interactions between these. In principle, any class can be modelled by a framework, and any framework can be abstracted as a class. However, as we have seen with books and borrowers, if there is nothing useful to say about the internals of an object, then no framework model is needed. Similarly, if there is nothing useful to say about a compound entity as an object, then no class definition is needed. Although we may *end up* with a hierarchy of class-framework abstraction, we do not set out to perform 'top-down decomposition'.

144   Chapter 12   Behavioural modelling

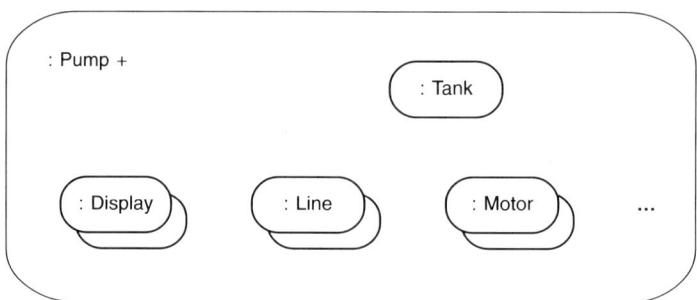

Figure 12.37 A Pump Group sharing a Tank

12.4.2 Overlapping

Suppose that in describing the behaviour of a single pump we had included the petrol storage tank as a component part. We might have done this because the behaviour of tanks seems to be at the same level of abstraction as the pump motor, the line, and so on. However, it may be that several pumps take their petrol from the *same* tank. So we might want to talk about 'the group of pumps that share the same tank', for example it might be the unit for maintenance scheduling.

Our 'pump group' entity therefore consists of one or more pumps, where the tank constituent of these pumps is 'shared', that is, it is the same tank object in each pump. We say that the pumps *overlap* on their tank subconstituent. A pump group could be drawn as in Figure 12.37. There are multiple 'stacked' boxes for instances of Motor, Display, and so on, but a single box for the instance of Tank.

We can also have overlapping between constituents of different kinds. This allows us to model situations where a single object plays two different parts in two (or more) different episodes of a history. For example, the same person might play two organisational roles: Finance Director on the Board of a Company, and Head of the corporate Planning Department in the same Company. This overlap between the Board and the Planning Department could be used to circumvent formal communications between them, since the person concerned could 'carry over' knowledge from one episode to the other. However, this might be neither necessary nor desirable.

12.4.3 Aggregation

There are some situations where we wish to represent constituent objects by an explicit *aggregation*. This typically occurs with collections of instances of the same class. For example, an order is made up of one or more order lines; each order line specifies a product and a quantity required. However, we also wish to identify and give attributes for the order as a whole. We could therefore regard an order as

12.4 Levels of abstraction

Figure 12.38 An Order as an Association

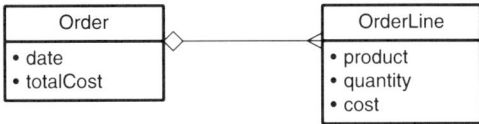

Figure 12.39 An Order as an Aggregation

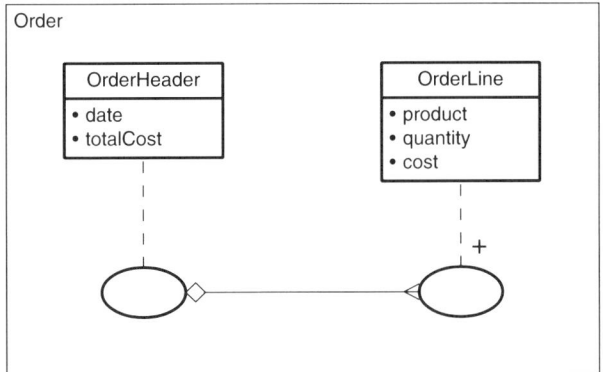

Figure 12.40 An Order Aggregation and its constituents

consisting of an order header, together with the order lines (Figure 12.38). The *date* attribute is essentially common to all the order lines, while the *totalCost* is derived by summing the cost attributes of each order line. However, an *OrderHeader* exists only to construct an order from a set of *OrderLines*.

We can make clear the special nature of the relationship by using an aggregation, rather than an ordinary association (Figure 12.39).

What we really have is two levels of abstraction: an *Order* and its constituents (Figure 12.40). Notice that aggregation is here expressed between constituent sets within a framework. The order header is referred to as the *aggregator*, and exists only in the context of an order. We usually want to express aggregation between classes, without explicitly identifying the higher level abstraction. A convenient shorthand is to give the aggregator class the name of the higher level abstraction, as in Figure 12.39.

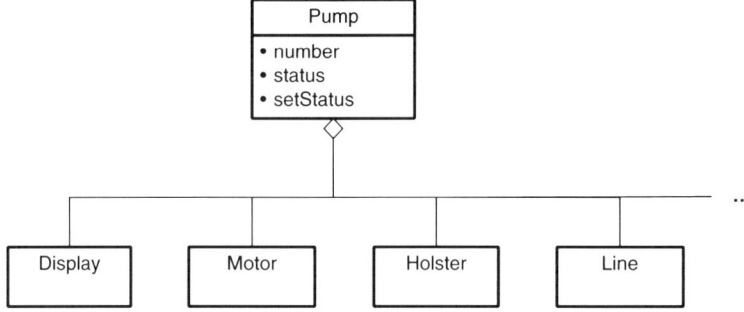

Figure 12.41 A Pump as an Aggregation

Collections containing instances of the same class are not the only kind of aggregation: we may also have aggregations containing disparate parts. In our petrol pump example, we might have decided that pumps needed to be identified explicitly by a pump-number, and had an overall operational status (in or out of service). Since these properties cannot obviously be delegated to any component parts, we could express them as properties of an aggregator class *Pump* (Figure 12.41).

12.5 The Class and Framework constructs

The previous sections have illustrated how behavioural modelling in ORCA is based around the dual concepts of class and framework. The examples given above have used the Beluga modelling language. We can now be more explicit about the modelling language constructs that correspond to these concepts.

A *class* consists of:
- a *class name*
- zero or more names of *parent classes* (classes of which this class is a specialisation)
- one or more *features* (attributes and operations)
- optionally, a statement of *invariant* properties

A *feature* is described by:
- a *feature name*
- a type *signature* (the type of an attribute, or the types of input parameters to an operation)
- a description of *meaning* (informal text, pseudo-code or formal specification)

- zero or more *facet* names, which label the feature (see Section 12.7 for more on facets)

A *framework* consists of:
- a *framework name*
- a description of *statics*
- a description of *class relationships*
- a description of *dynamics*
- optionally, a statement of *invariant properties*

Framework *static structure* covers
- declarations of *constituent sets* (single objects or multi-object groups)
- optionally, a statement of *overlaps* (shared subconstituents)
- optionally, *aggregation* relationships between constituents

Class relationships include:
- *association* relationships between classes
- *interaction* relationships between classes

Framework *dynamic behaviour* is an expression containing frameworks that describe events, episodes and processes, combined using various constructs: associate events, interact events, composition, choice, and so on.

The full syntax of the Beluga modelling language, together with its textual and diagrammatic forms, is given in Appendix C.

12.6 Structural frameworks and temporal frameworks

In Section 12.4 we have a *Pump* framework modelling the internal static structure of a petrol pump. On the other hand, in Section 12.1 we have a *Loan* framework modelling a behavioural episode. These are expressed using the same language construct—the framework—but one provides a structural view of behaviour, while the other provides a temporal view.

However, if we go back to our original view of behaviour as having both 'object space' and 'time' dimensions, it is apparent that we can subdivide behaviour in either dimension (Figure 12.42). Since any two-dimensional chunk of behaviour can be characterised by a framework, we can use this construct for both structural and temporal views. Since 'structures' can have a finite lifetime, and 'episodes' can be complex and long-running, there is no fundamental distinction between the two.

On the other hand, in a given framework, there may be subframeworks that model the classes of constituents, and subframeworks that appear in the description of dynamics. It is convenient to regard the former as 'structural' frameworks,

148   Chapter 12   Behavioural modelling

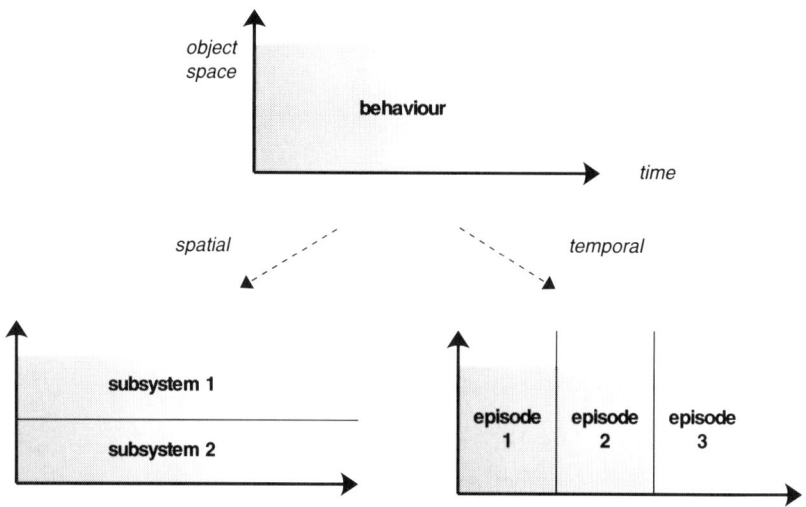

Figure 12.42 Dimensions of behaviour

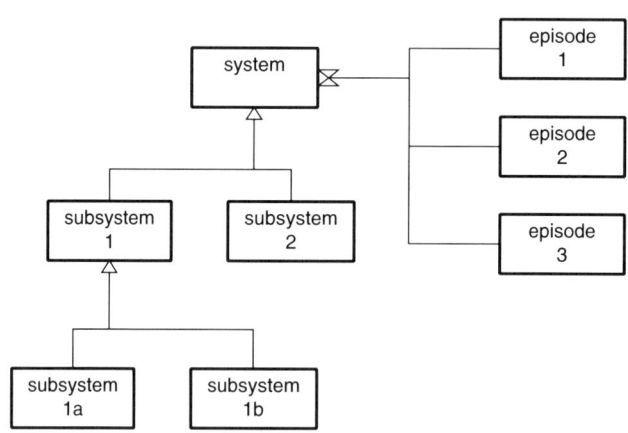

Figure 12.43 Structural and temporal decompositions

and the latter as 'temporal' frameworks. One way of visualising these abstraction relationships is as decompositions in 'object space' and time dimensions (Figure 12.43). The structural decomposition proceeds top-to-bottom, and is indicated by the 'pyramid' symbol; the event decomposition proceeds left-to-right, and is indicated by the 'hour-glass' symbol. 'Abstraction maps' of this kind can be useful in keeping track of frameworks and classes at different levels of abstraction.

12.7 Behaviours and Services

Purposive models provide descriptions of *services* relied on and guaranteed by co-operating roles. The system's behaviour should realise the services. It follows that aspects of behaviour concern particular services:

- Every object, and thus every class, exists to realise one or more services.
- Features defined by a class exist to realise one or more services.
- Frameworks describing kinds of dynamic behaviour exist to realise one or more services.

For example, suppose that our Lending Library is intended to provide two services:

- managing the loaning of books
- providing reference and research facilities

All the classes in our behavioural model should be relevant to one or both of these services. The *Book* class is clearly relevant to both services, whereas the subclasses *LoanableBook* and *ReferenceBook* are relevant to the first and second services respectively.

The features of the *Borrower* class (perhaps more accurately called *LibraryMember*) could be labelled according to their relevance, using facet names such as *loaning* and *reference*. Some features, such as basic attributes and features to do with library membership, might belong to both facets.

The dynamics of the library might be modelled as a composition of loan-related transactions, together with a composition of reference-related transactions. Library members would be involved in both kinds of transactions, but using different (possibly overlapping) subsets of their features. The features in these subsets should be those labelled with the appropriate facet name in the class definition.

As the example outlined above indicates, the 'mapping' between purposive and behavioural models is by no means a simple matter. It is best treated as a heuristic for analysis: for any aspect of a behavioural model, it should be possible to ask 'to which of the services described in the purposive model is this relevant?'. It should be remembered throughout analysis that systems have purpose as well as behaviour, and that as analysts we are concerned to tailor each to the other.

Chapter 13
Do's and don'ts

13.1 Introduction

The heuristics—guidelines and 'rules of thumb'—given in this chapter provide some hints on how to apply ORCA, and on how *not* to apply it. They should not be applied blindly and rigidly, but breaking the rules in a particular situation should be justified.

The heuristics are grouped under headings indicating where they might be applicable. However, some fit under more than one heading (for example, hints on naming subclasses are relevant both to 'Classification' and to 'Naming').

13.2 Process

Don't go for completeness

Completeness is unattainable. Beware of 'pan-galactic' diagrams and models.

Don't use Grampus for behaviour

If the behaviour has become more interesting than the co-operations between roles, it is time to stop doing purposive modelling with Grampus, and instead do some behavioural modelling with Beluga.

Talk to everybody

Anyone involved with the client's world has something useful to say. Don't assume you know everything, and don't assume the client does either.

Expect to iterate

...and don't be afraid to do so (even though it is impossible to manage...).

13.3 Modelling

Model enough of the environment
Question, and widen, any preset analysis boundaries (but be aware that this may present some 'political' problems).

Systems are not their organisational structure
'Organograms' may not reflect the actual *purposive* entities.

Grampus co-operations identify Beluga classes
Use the vocabulary of Grampus service descriptions to help identify behavioural entities (so, underline the 'Grampus nouns')

Don't automate the Old World; rather, design a New World
Many 'objects' in the Old World, especially pieces of paper, may cease to be necessary in the New World. Check their *purpose*.

Don't build an IT system
...except as a last resort.

Quantify Grampus qualifiers
Try to be more specific than 'in a timely manner', 'reliably', ..., especially when they occur at both ends of a co-operation, with potentially different meanings.

...but don't invent spurious quantities
If the quantity is 'nearly all the time', don't just guess that this means 'more than 50%', or '99%', or '99.99%', or.... Call it 'nearly all the time', until you know the actual proportion. Question where quantitative requirements ('respond in less than 3.14159 seconds') come from.

Document assumptions
...for example by using Grampus justifications.

13.4 Diagrams

Small is beautiful
Aim for no more than '7±2' major elements per diagram (reliances and guarantees per role, classes per framework, ...).

A rough diagram that is right is better than a pretty one that is wrong

Don't use a diagram where words will do

Not all diagrams are worth a thousand words... but where diagrams are expressive, use them. Text is sequential, so it is good for expressing sequential structures. Indented lists of names can be used for tree-structures (such as organisational structures). Box-and-line diagrams are good for networks of relationships.

13.5 Abstraction

Identify abstract frameworks

... and customise them for this analysis.

It's too detailed

You are working at too low a level of abstraction. Don't overcomplicate things; abstract to find the essence and commonalities. 'Make it as simple as possible.'

But it's more complicated than that...

You are working at too high a level of abstraction. Don't oversimplify; understand the detailed consequences. 'Make it as simple as possible, *but no simpler.*'

All behaviour is concurrent until proven sequential

Distinguish the necessary dependencies between bits of behaviour from sequential implementations in the Old World. Don't build big dynamic models.

13.6 Classification and specialisation

Don't classify

Don't distinguish things that don't need to be distinguished at the analysis level. Don't set out to do a 'taxonomy'. Three levels to a generalisation/specialisation hierarchy is probably plenty (for analysis).

Use behaviour to guide specialisation

Specialisation of classes should capture either behavioural extension ('more' behaviour than the parent) or behavioural alternatives (between sibling classes).

Don't use multiple inheritance

Multiple inheritance is a useful coding technique for 'mixin' functionality—but this is not an analysis issue. At the analysis level, its use could well indicate excessive classification.

13.7 Naming

Subclass names can be noun phrases

Any subclass should be nameable as a qualified superclass name, even if this is not the name eventually used. For example, a Dog *could* be called a Barking-Mammal. If you can't name the subclass this way, have you really got a subclass, or something else?

Facet names can be gerunds

... that is, words ending in '-ing', such as 'printing', 'registering', 'enrolling', ...

Guarantees are active, and reliances are passive phrases

Guarantees should be expressed as an active qualified phrase: 'do x', 'weave name tapes'. Reliances are passive: 'it is the case that x is done', 'name tapes are woven'.

Service descriptions should have qualifiers

'weave name tapes *reliably*', 'payment is handled *promptly*', ... Qualifiers can be used as the hooks to capture so-called 'non-functional' requirements.

Part IV

Tailoring the Process

Chapter 14

Introduction to tailoring

14.1 Why do we need to tailor the process?

It is in the nature of analysis that we should be capable of performing it in any situation—'any time, any place, anywhere'. No requirements can be made about the analysis situation—the analysis process must be tailored to fit it.

ORCA's approach is to provide a basic analysis process which can be used as the starting point for a tailoring activity. This explicitly acknowledges what inevitably happens in practice.

ORCA's Basic Process is described in detail in Part II; the subsequent chapters in Part IV illustrate different tailorings for different analysis situations. These examples are not intended to provide a comprehensive catalogue of process tailorings, but should be sufficient to indicate the range of variation.

It is worth pointing out that an analyst cannot be expected to make a decision about the appropriate kind of process without first gaining some familiarity with the analysis situation. One of the aims of the Preliminary Analysis activity is to allow an appropriate process to be determined.

14.2 Overview of tailorings

ORCA's Basic Process is shown in Figure 4.1. The tailorings described in the subsequent chapters of Part IV are summarised using the same form of diagram, but with differently emphasised activities and different dependencies. The tailorings emphasise some activities (shown as emboldened ellipses), and de-emphasise others (as dashed ellipses).

The following chapters describe a range of example tailored processes. The first three are based around real case studies (although some details have been exaggerated or invented for our own expository purposes), with well-defined, common pathologies:

15 Organising the Organisation—National Parks. Internal pathology: the organisation is disorganised.

16 Shaking Up the Business—Just in Time. The current structure is blocking desired growth.

17 A New Purpose in Life—The Paperless Map. External pathology: the organisation must change to track a changing environment.

The remaining examples are used to illustrate certain parts of the process not highlighted by the NIMWeC case study:

18 The Missing Old World—Spreadsheets and Telephony. When developing a new product rather than enhancing an old system.

19 Nothing New Under the Sun—a Lending Library. How reuse fits into the process.

20 No-one to Talk To—Ahab applied to NIMWeC. When the sources for 'information gathering' are restricted to impenetrable documentation.

Chapter 15

Organising the organisation—National Parks

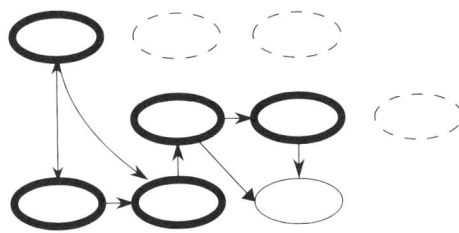

15.1 Analysis parameters

Typical scenario

- The client's world is a large, distributed organisation with multiple independent units. The structure of the organisation may not be well defined.
- The client is initially aware of a general malaise in the organisation, manifested as a variety of 'low level', seemingly unconnected problems.
- The client initially has no definite proposals for software development.
- Typically there are multiple sources of information (for example, representatives of different organisational units, other interested parties).
- The services involved may often be ill-defined; areas of activity may cut across organisational structure.

Aims of the analysis

- The overall aim of the analysis is diagnostic—to identify a pathology that explains the various problems and to suggest courses of remedial action.
- Specifically, the analyst needs to produce an Old World purposive model that can be used as the medium for problem analysis, and a statement of pathology resulting from this analysis.
- The analyst should produce recommendations for remedial action. Where changes are organisational or procedural, a modified model may be produced.
- Where appropriate, specific target areas for the development of IT support can be targetted and the scope of further analysis outlined.
- The results of analysis may be needed to aid the client in making cost/benefit decisions on candidate proposals.

15.2 Preliminary analysis

Preliminary analysis is important in this kind of process, establishing the scope and motivation for the analysis and providing an initial orientation.

◇ *Determine the reason for analysis (external stimuli, internal problems). Record and structure informal statements problems.*

Our client is the Gingandan Association of National Parks (GANP)—an association of locally administered National Parks. GANP is undertaking a thorough review of its organisation and procedures. Numerous low level problems are reducing its effectiveness; complaints have been received from the local Parks about the usefulness of the national office. Initial impressions are that this is due to the informal structure of the Association, and poorly defined or *ad hoc* procedures. Installation of IT support in appropriate areas is being considered, but there is no overall strategy for this.

◇ *Determine scope of analysis (what is and isn't open to consideration) and external constraints.*

The analysis is to consider the organisation and procedures of GANP. The overall purpose of GANP (that is, the promotion and co-ordination of the National Parks) is not open to change.

◇ *Gain an overview of the domain. Identify interested parties, objectives, issues, potential conflicts, constraints.*

Organisations such as GANP are not rigidly delimited or structured. It is therefore important to obtain an initial idea of what is relevant to the analysis. In addition, the analysis activity might well be constrained or complicated by 'political' considerations: the various interested parties may have different, and not necessarily compatible, objectives (beware!).

A rich picture for GANP is shown in Figure 15.1.

◇ *Arrange information gathering (sources, access, where and when).*

For organisations such as GANP, it is important to gather information from as many viewpoints as possible. As we shall see, many of the problems arise from incompatibilities between different components of the organisation. Consequently, information gathering may be a substantial activity and need a certain amount of planning.

15.2 *Preliminary analysis* 161

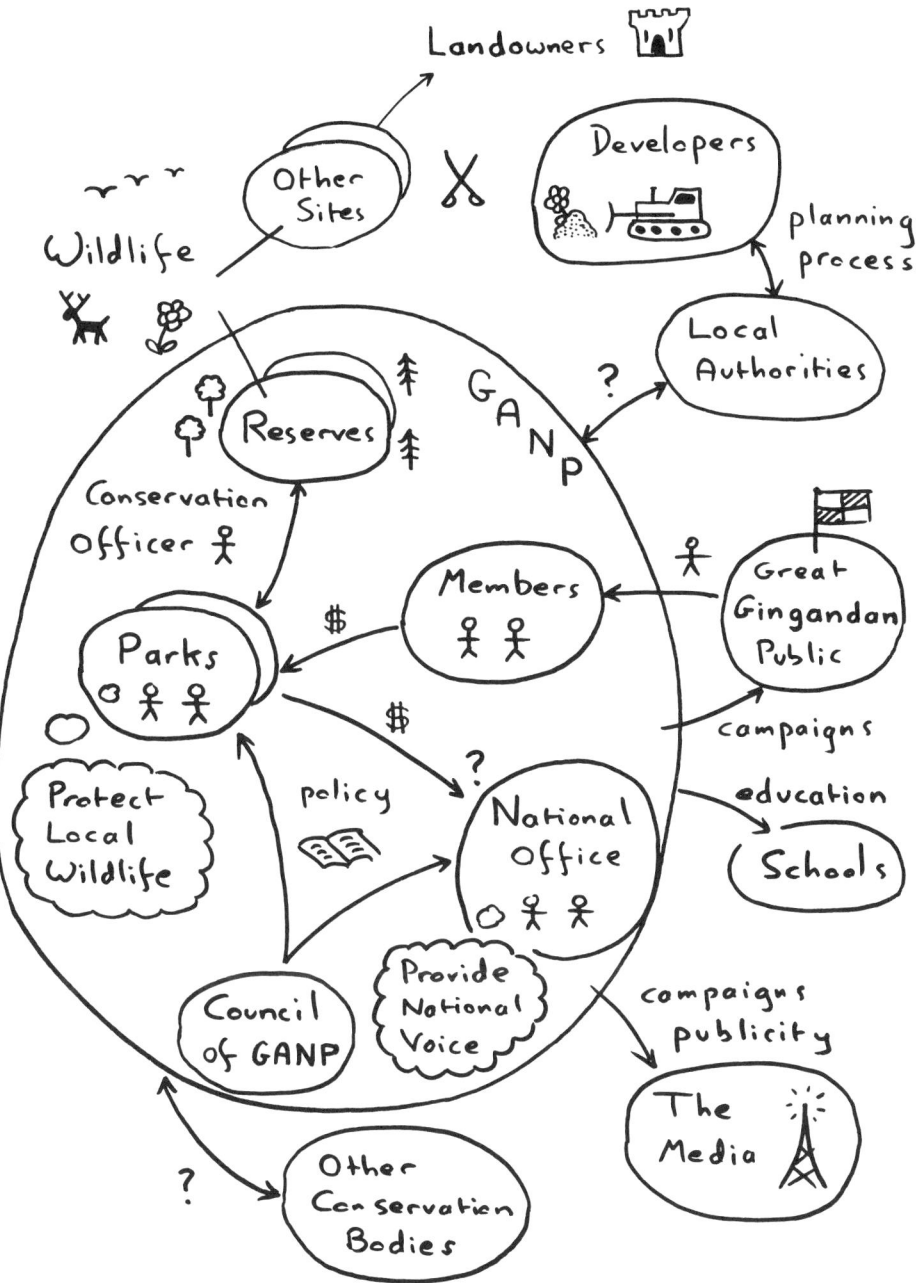

Figure 15.1 GANP Rich Picture

15.3 Modelling the Old World

◇ *Produce a role decomposition—attempt to identify purposive entities; the existing organisational structure and activity areas may both be considered.*

The top level organisation of GANP consists of a single National Office, numerous local Parks (usually at provincial level), and a Governing Council. We can also take an activity oriented view of GANP. The main activity areas are:

- Campaigning
- Conservation
- Education
- Membership
- Funding
- Policy & Management

We can view GANP as a set of 'partnerships' between the Parks and the National Office—one partnership for each activity area. The National Office acts as coordinator and provider of centralised services for each partnership. Each Park is involved in all areas of activity, as is the National Office, but the Council is involved only in the policy area.

Within the Parks, the conservation activity is further subdivided into Wildlife Resource Management, Development Control, and Conservation Information.

There are many ways to represent this sort of organisational structure. We could use Beluga, an organisation chart, or a matrix diagram, for example. In this case, since the structure is matrix-like, we choose the latter (Figure 15.2). This gives us roles at various levels: GANP as a whole, the organisational units (National Office, Park 1, ...), the activity area partnerships (Campaigning, Conservation, ...), the intersection of organisational units and activity areas (National Office Campaigning, Park Education, ...), and the three subroles of Park Conservation.

◇ *Use the role decomposition as the basis of purposive modelling; examine guarantee–reliance relations (co-operations and delegations) throughout the organisation.*

Delegations should exist between any role and its subroles, for example between GANP and the Parks, and between a Park and its Park Campaigning subrole.

Co-operations may exist between any roles in the same structural layer, for example between the National Office and the Parks (as a group), between Park Conservation and National Office Conservation within single Conservation partnership, and between Park Conservation and Park Education within a single Park.

◇ *Check that all services guaranteed or relied on at one level map to guarantees or reliances at a lower level.*

What are the top level guarantees–reliances of GANP? How are these delegated to the Parks, the National Office and the Council (or to the various activity areas)? Put another way, how do the various subroles of GANP contribute to its

15.3 Modelling the Old World   163

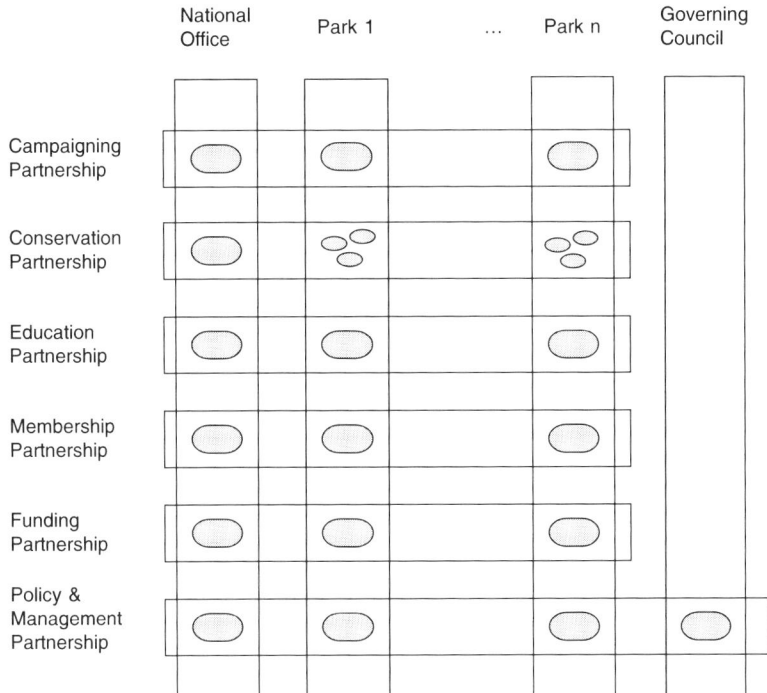

Figure 15.2 GANP organisational structure

overall purpose? This may be difficult to establish, where the top level services are quite vaguely described. For example, a top level service might be to 'raise public awareness of conservation issues'; many (perhaps all) subroles contribute to this in different ways.

◇ *Check that all services guaranteed or relied on by roles in a layer are involved either in co-operations with roles at the same level or in delegations from a higher level.*

In any large organisation it is possible that there is a certain amount of 'dead wood'; services are provided that are no longer required either by other roles at the same level or by higher level roles. This is commonly the case with reporting services ('who needs those regular monthly reports?', 'what do they use the information for?').

◇ *Check that co-operations between roles are okay: that all extrinsic service descriptions can be matched to intrinsic ones.*

This is a key activity in combining multiple viewpoints into a single purposive model. The process cannot be mechanical, since different viewpoints may contain different (informal) descriptions of the same service. Examples of faulty co-operations are given in the next section.

◇ *Examine the behaviour that realises each lowest level co-operation, in order to determine its effectiveness—does it actually work? Consider how effectiveness can be measured.*

Even if co-operations are purposively okay, they may still be behaviourally problematic: a service may not be realised effectively; there may be failure of co-ordination between client and server; there may be inadequate resources to provide the service. Examples of behavioural faults are given in the next section.

◇ *Look for 'hidden' or informal interactions.*

Do Parks have co-operations amongst themselves, or only with the National Office? Is there any co-ordination of activities or sharing of resources that takes place purely between Parks? This is an important area to look at, since it may be the case that the explicit co-operations (between Parks and the National Office) seem to work only because there are hidden interactions between individual Parks.

15.4 Determining the system pathology

It is typical of this kind of analysis that the organisation has a compound pathology; there is no single cause for the various problems. In the GANP, there are a variety of subpathologies, and these are illustrated below.

◇ *Group and classify problems.*

An example of *purposive mismatch* concerns the Park Education activity area. Various other components of a Park assume that Park Education provides training services to staff and volunteers. In fact, Park Education's services are directed primarily at educating the general public in conservation matters; this is a responsibility delegated from the Education Partnership and ultimately from the GANP as a whole. The pathology here is that Park Education is not providing the services required by its peer components within a Park (Figure 15.3). This pathology is a result of the dual structuring of the GANP system: organisational and activity-area. Park Education is providing those services that are implied by the activity-area structuring, but not those required by the organisational context.

An example of *faulty realisation* ('the spirit is willing but the flesh is weak') occurs with the generation of revenue from members. Park Funding requires Park Membership to provide revenue from members; this requirement is accepted by Park Membership (Figure 15.4). However, the activities performed by Park Membership fail to satisfy this requirement—not enough revenue is generated. Either the current activities need to be pursued with greater vigour or new activities need to be considered.

15.4 Determining the system pathology

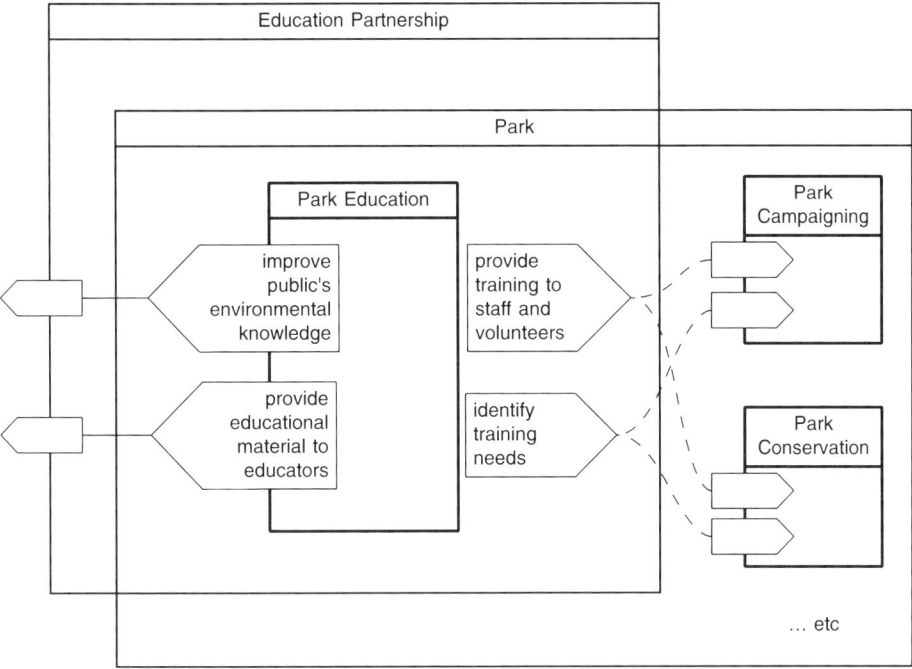

Figure 15.3 GANP pathology

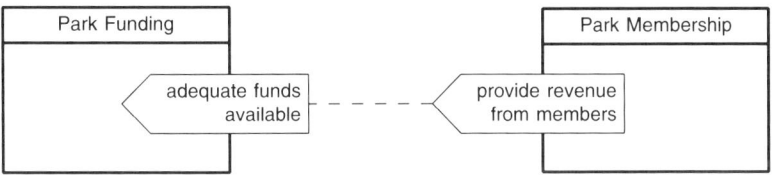

Figure 15.4 Park Funding's revenue

A *co-ordination problem* exists with regard to the provision of management and operational information. This is required by Park Policy & Management, and is provided by the other Park components (Conservation, Membership, etc.) (Figure 15.5). However, the information is not provided at the right time, in the right form, or at the right level of detail. There is also the issue of consistency between the various providers of the information—Park Policy & Management wants the information to be provided at the same time, in the same form, and so on.

An example of *inadequate resourcing* occurs with the provision of survey information by Conservation Information to Development Control. The latter compo-

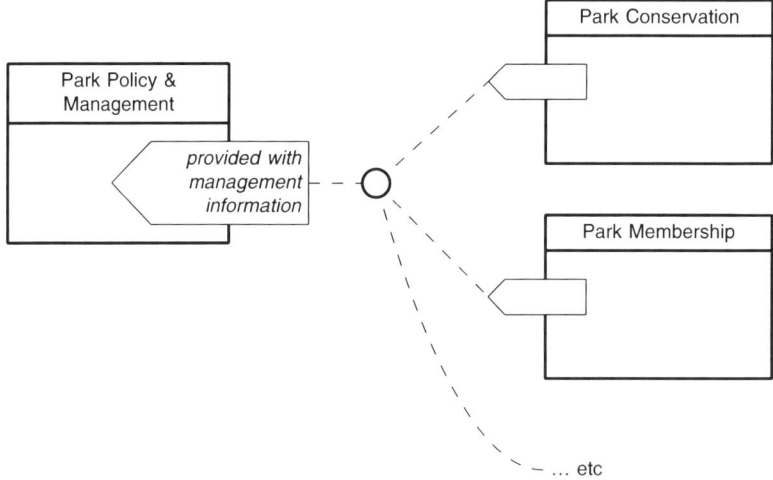

Figure 15.5 Management information

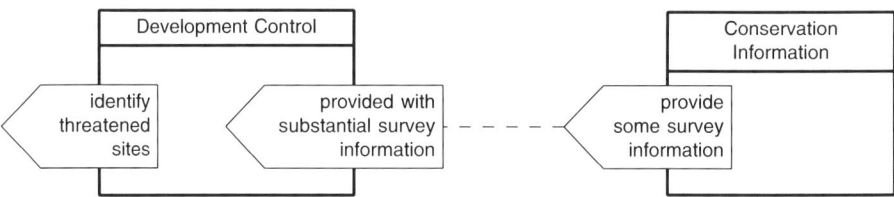

Figure 15.6 Survey information

nent needs information about ecologically important sites within the Park's area, so that development plans affecting such sites can be identified. In order to produce this information, Conservation Information carries out an ongoing survey of the Park's area using specialist staff and volunteers (Figure 15.6). However, the task is a substantial one and resources are generally insufficient. Consequently, the survey information is neither comprehensive enough nor up-to-date enough. Secondary sources of information (local authority records, the general public, etc.) are used to make up for this deficiency, but this then creates a co-ordination problem in that the information is then inconsistent in form and content.

An *inertia problem* occurs when an interaction appears to be okay both purposively and behaviourally, but nobody actually instigates the interaction. As an example of this, National Office Campaigns should provide Park Campaigns with ideas and materials for local campaigns (Figure 15.7). Although the procedures are in place, the Parks do not typically request help with local campaigns. Why?

In all cases where there is a problem, there is also potentially a *monitoring*

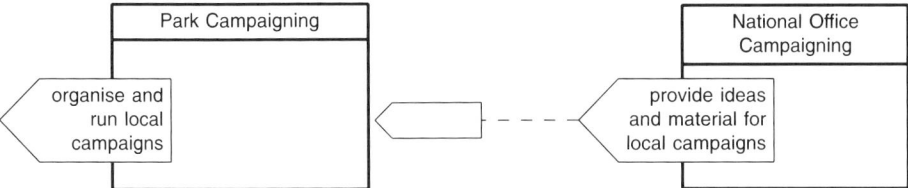

Figure 15.7 Campaign information

and control problem. The parties to a co-operation or some third party should be responsible for monitoring the effectiveness of the co-operation and instigating some corrective action when necessary. For example, how does Park Policy & Management tell the other Park components what information it needs? How does Park Membership know whether it is raising enough money?

◇ *Assess the criticality and tractability of problems.*

Given this compound pathology, which aspects of it have the most serious effect on the organisation? Which problems need addressing urgently, and which are of secondary importance? How easy is it to solve particular problems? In this case, we might decide that the lack of resources for surveying is linked to inadequate fund-raising, and that this is the most significant problem, but also difficult to solve. On the other hand, the co-ordination problem regarding the provision of management and operational information could be solved relatively easily by a mixture of better-defined procedures and IT support.

15.5 Drawing conclusions—the Analysis Report

The analysis models themselves might be delivered to the client, but the quantity is often too great for this to be feasible. It is more likely that the analysis models are used as support for an Analysis Report. This report should typically address the following points.

◇ *Identify and assess the most significant problems.*

This is essentially the output from determining the pathology.

◇ *Make specific recommendations for remedial action.*

These recommendations could include any of the following:
- restructuring of the organisation
- reallocation of responsibilities
- changes in procedures
- explicit definition of responsibilities or procedures

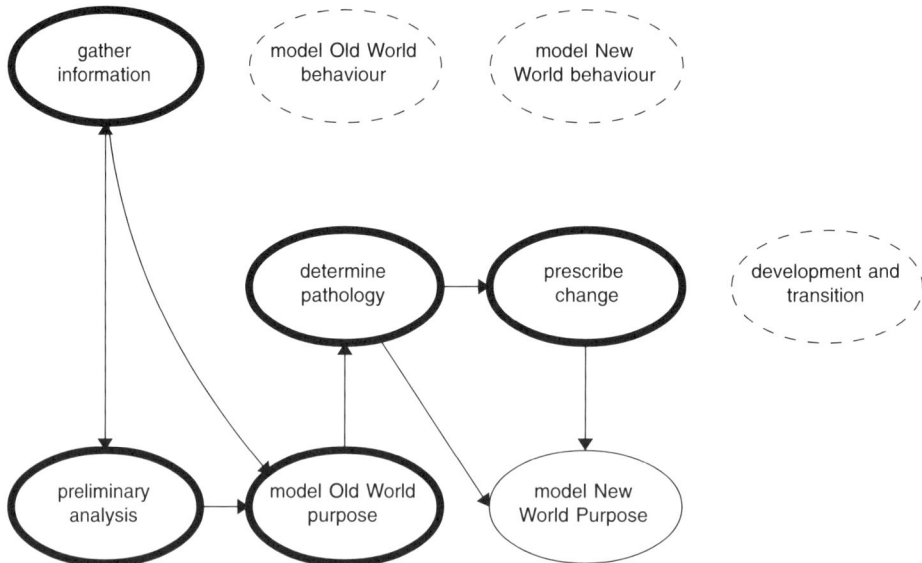

Figure 15.8 The tailored process

- increase in resources
- staff training
- capital purchase
- software development and installation

◇ *Prioritise and (where possible) cost the proposed remedial actions.*

◇ *Where specific areas for the development of IT systems have been identified, outline the nature of the analysis necessary to determine the behavioural requirements.*

In our GANP example, we might conclude that each Park needs an IT system to handle management and operational information. Further analysis would then be needed to determine the detailed requirements of such an IT system—the type of information to be recorded, the access and storage requirements, and so on.

◇ *Where proposed software development is extensive, set out a development and installation strategy (for example, incremental development, parallel running of old and new IT systems).*

Chapter 16

Shaking up the business— Just in Time

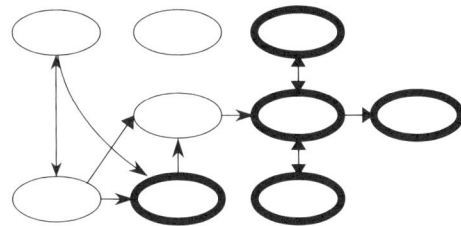

16.1 Analysis parameters

Typical scenario

- The client's world is a large business organisation with multiple operating units.
- The client has explicit goals that are providing the stimulus for change (for example, 'double the production').
- Actions necessary to achieve the goals are being blocked by limitations of the existing organisation.
- The client is prepared to make organisational changes to the business.
- The client may have existing software systems; these may or may not need to be retained.

Aims of the analysis

- The overall aim of the analysis is to re-engineer the business process so as to remove limitations that are preventing desired changes.
- Specifically, the analyst needs to produce one or more New World models that describe modified circumstances, together with justification for their effectiveness.
- The analyst should identify specific changes to be made to the organisation; these changes may involve the development of IT support.
- The analyst may need to produce an initial specification of IT support, where this is appropriate: that is, of information, processes, external communications, etc.
- A strategy for hand-over to the target IT system will need to be produced.

16.2 Preliminary analysis

◇ *Determine the reasons for analysis (external changes, internal problems). Record and structure informal statements of problems.*

Our client is Birfami (Imbirfa) Motors, BIM, a large car manufacturer. The manufacturing operation at one particular plant is under consideration. The external stimulus for change is an increase in demand for BIM's products. To exploit this, BIM wishes to increase the volume of production at the given plant. This would entail increasing the scale of the production facilities. However, the physical extent of the plant is limited, and it is not feasible to split operations across several sites. A large area of the current site is used by the stock control department, which maintains stocks of parts and materials, and subjects incoming goods to quality tests. BIM believes that a solution must lie in reducing the spatial requirements of this department, allowing the production facilities to be extended.

◇ *Determine the scope of analysis (what is and isn't open to consideration), and external constraints.*

The analysis is to consider the organisation and procedures of the business; it is not to consider marketing or financial issues.

◇ *Determine scope for change within the business.*

Car production is the primary activity of the business, and must remain so, although minor changes to the operation can be countenanced. The stock control function is ancillary, and can be modified as necessary. Relationships with suppliers of parts and materials are potentially changeable; BIM is generally in a strong position with respect to its suppliers.

◇ *Arrange information gathering (sources, access, where and when).*

The primary sources of information are senior personnel—heads of departments, etc. We may need some 'observation in the field', in order to gain a full understanding of current mechanisms and constraints.

◇ *Model the organisational structure of the business.*

BIM is organised into six departments
- Production
- Stock Control
- Planning
- Sales & Marketing
- Personnel
- Accounts

The Production department is organised into numerous production units. Each unit carries out a sequence of production runs, according to an overall production schedule. A production run involves doing a specified manufacturing operation for a specified period. The Stock Control department contains three divisions:

- Testing
- Warehousing
- Orders & Invoicing

The Testing division is responsible for quality control of incoming parts and materials. Warehousing holds stocks of parts and materials for use by the Production department. Planning, Sales & Marketing, Personnel and Accounts departments have their own divisional structures (which we omit for brevity).

At this stage, we don't know which parts of the business are going to be relevant to the analysis.

16.3 Describing the Old World

◇ *Focus the analysis on particular areas of the world.*

BIM has outlined the basic problem: to free some proportion of the space currently taken up by the stock control function. It therefore seems sensible to focus on Stock Control, and 'work outwards' from there.

◇ *Use purposive modelling to understand co-operations.*

The organisation of BIM has been designed, ar at least evolved over time, to operate effectively. Our current analysis is essentially a re-design exercise for the business. Thus, we can assume that our client has control over the roles of the organisational divisions. It is therefore reasonable to take the organisational divisions as roles within a purposive model of the system. This is in contrast to the role decomposition for GANP (Chapter 15), for which the organisational structure was not sufficient.

In order to understand BIM, we need to look at how the Stock Control department co-operates with other departments. This is summarised in Figure 16.1.

The Production department issues requests to Stock Control for parts and materials needed in the production process. These requests specify the kind of item, quantity, the production unit requiring the items, and when the items are required. Short-term requests (for example, 'As Soon As Possible') are generated as parts and materials are consumed by the production process. Long-term requests (for example, 'In Six Weeks Time') reflect changes in the output products, as indicated by the production schedules generated by Planning (for example, a new colour of car is to be produced after a certain date; stocks of the new paint must be available at the appropriate time).

The requests for parts and materials are handled by Warehousing. Short-term requests are satisfied from current stocks; long-term requests are passed on to Orders & Invoicing. Warehousing is responsible for moving items to the production

172 Chapter 16 Shaking up the business—Just in Time

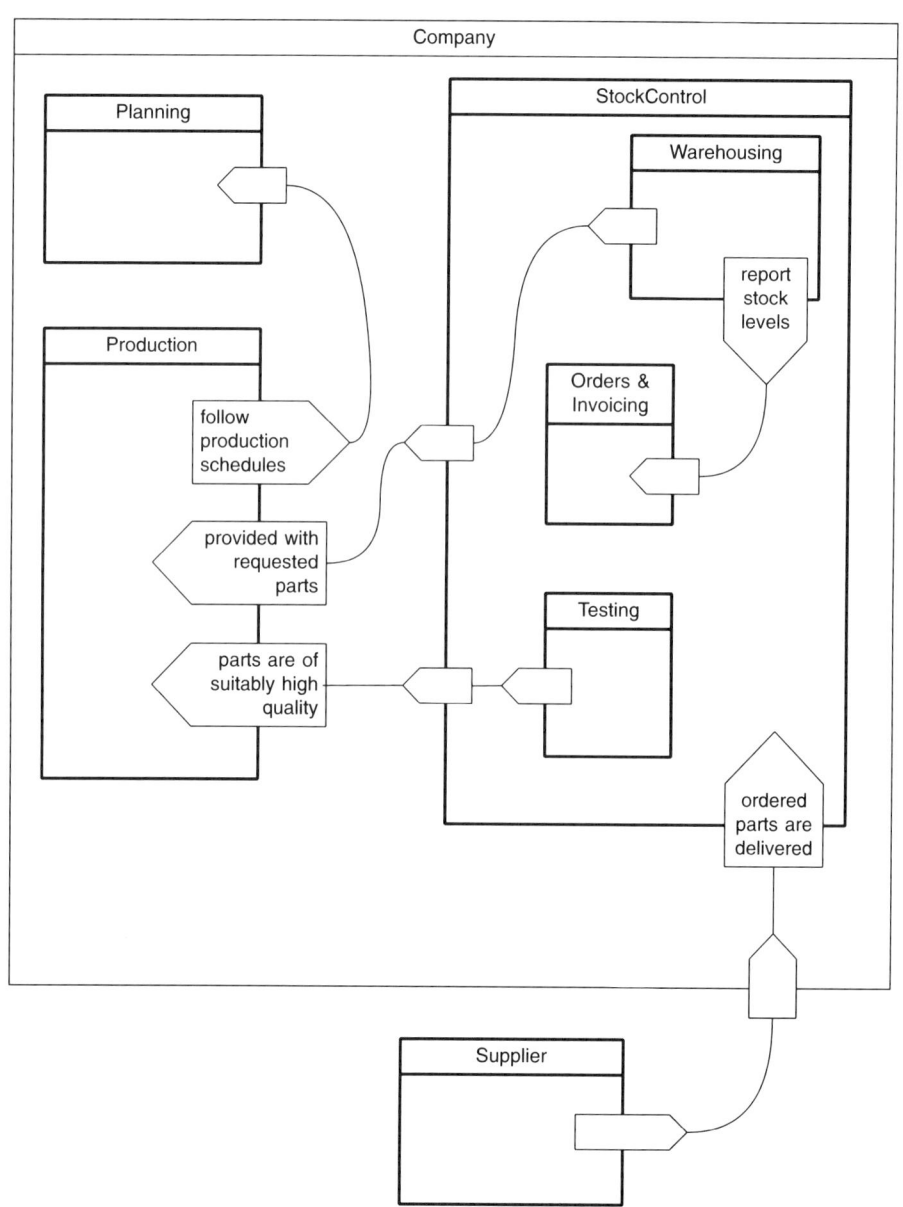

Figure 16.1 Old World purposive model

unit that originated a request. In addition, Warehousing provides periodic stock level reports to Orders & Invoicing, who generate orders as necessary to maintain stock levels above minimum levels.

The quality of parts and materials is ensured by Testing. Delivered goods are received and checked by Testing prior to being passed to Warehousing for storage.

There is an extrinsic reliance on Suppliers to deliver ordered parts. The degree to which Suppliers provide this service is not under the direct control of BIM (but it may influence it, by threatening to change Supplier).

Some departments within BIM (for example, Personnel) do not appear in the model shown above because they do not have a direct impact on the behaviour of Stock Control.

◇ *Identify and record current values for suitable 'critical success factors'.*

Since we are interested in the use of physical space, it is worth recording space attributes of the three subroles that form Stock Control. In this case, both Warehousing (storage area) and Testing (many people) use large amounts of space, whereas Orders & Invoicing consists of relatively few people in a small office unit. In principle, we could record other attributes, such as staff numbers and cost. When we come to specify a New World, we can then make explicit the quantitative improvement that we expect to achieve. In this case, we need to determine what constitutes a significant increase in the space available for the production facilities.

◇ *Where necessary, use behavioural modelling in order to clarify the informal understanding of services.*

Notice that the model given above is a purposive, not a behavioural, model. There are implied physical 'flows' between the various roles: parts requests, parts and materials, production schedules, stock level reports, and so on. However, we should be very cautious about behavioural modelling at this stage. We don't yet know which parts of the world are going to be changed, or how. Behavioural modelling of interactions that are going to be different in the New World (or absent entirely) is generally not cost-effective. We may need to do a certain amount of behavioural modelling in order to clarify our informal understanding of services, but this should be done judiciously.

16.4 Determining the Pathology

◇ *Identify the properties of the Old World that are blocking the desired changes (or if these were identified in the Preliminary Analysis, confirm that these are indeed the culprits).*

The client's world is not pathological as such. It works well enough, assuming that the interaction with suppliers is satisfactory. The problem is to do with change:

a desired expansion of the production facilities is being blocked by the physical space requirements of Stock Control. In this case, there is no doubt that this is the characteristic that has to be changed. We now have to ask: how can the overall business be re-engineered so as to alter the limiting characteristics?

Notice that this scenario is different from one in which the part of the world being analysed doesn't work properly because of internal problems. In this latter case, the task of the analyst is to specify a New World that remedies the pathology (see Chapter 15 for an example).

16.5 Specifying a New World

◇ *Identify key 'variables' in the Old World model.*

Creating a New World is not a mechanical process. In the kind of analysis being illustrated in this chapter, finding a solution often needs a mixture of 'lateral thinking' and 'seen this one before' experience. In our car production example, there are a number of important insights:

- BIM is in a very strong position with respect to its Suppliers (typically, small firms making parts to BIM's specification).
- If BIM mandates Suppliers to provide the quality control service, the need for the Testing division would be removed.
- Warehousing currently maintains large stocks of parts and materials because (a) delivery times of ordered items are highly variable, and (b) there is no monitoring and prediction of consumption by Production. If these two problems can be addressed, then Warehousing would not be needed.
- BIM could mandate Suppliers to deliver ordered items to schedule (with 'guaranteed timeliness').

◇ *Produce (one or more) purposive New World models, equivalent in scope to the purposive Old World model.*

One possible New World is described in Figure 16.2.

Stock Control is still responsible for providing Production with parts and materials. However, rather than Production generating explicit requests, Stock Control has a new Monitoring & Prediction subrole, which provides Orders & Invoicing with detailed ordering schedules; as well as kind and quantity of item, these specify the required delivery date and the destination production unit. Ordering schedules are generated on the basis of consumption statistics from Production (indicating current usage) and product updates from Planning (indicating planned introduction of new products, deletion of products, or changes in product specifications).

In the Old World, Testing received items on delivery from the Suppliers; Warehousing was responsible for moving items to production units. These two services need to be retained in the New World, but can be separated from Stock Control

16.5 Specifying a New World

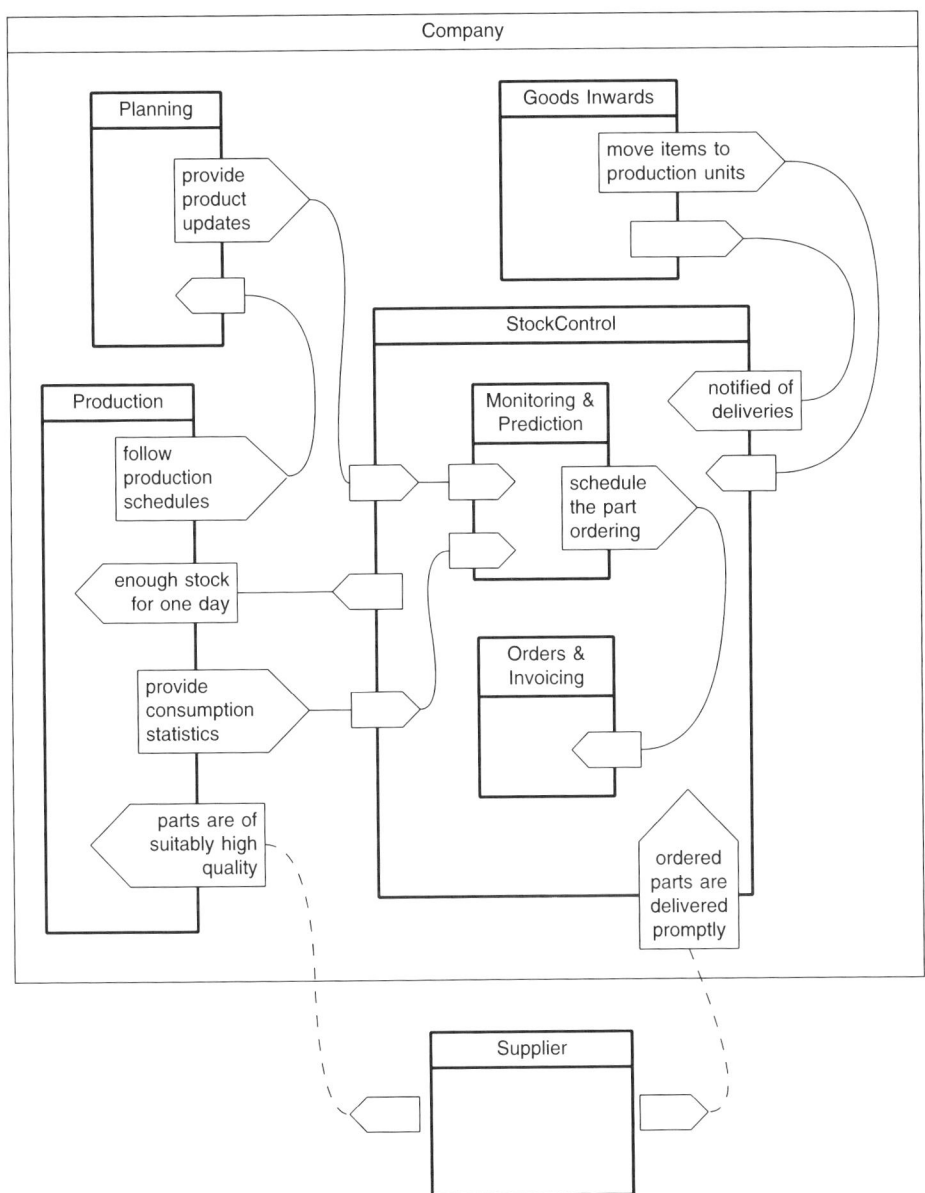

Figure 16.2 New World purposive model

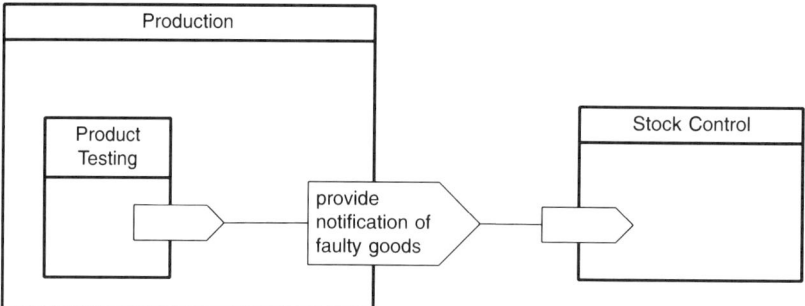

Figure 16.3 Product testing

as a new role, Goods Inwards. Note that the physical parts and materials pass from Suppliers to Goods Inwards to Production; Stock Control no longer handles physical items. (In the original implementation of 'Just in Time Manufacturing', Suppliers delivered directly to production units.)

◇ *Determine how extrinsic reliances are monitored, and what the consequences of service failure are.*

There are now two extrinsic reliances on Suppliers: timely delivery of ordered items and quality control. How is the provision of these services to be monitored? In the first case, Goods Inwards notifies Stock Control of deliveries; the delivery date can be compared with the required delivery date on the ordering schedule. Suitable action can be taken with Suppliers who do not perform adequately (they will have been warned!).

In the second case, Production needs to increase the importance of product testing, since product defects may be due to faulty parts or materials, as well as to production errors. There needs to be some kind of feedback to Stock Control about the quality of parts and materials (Figure 16.3).

16.6 Prescribing Change

◇ *Identify the principal changes.*
- Get rid of Warehousing and Testing divisions from Stock Control.
- Move the receipt and distribution of items into a new Goods Inwards division.
- Implement a Monitoring & Prediction division within Stock Control.
- Modify Production so that consumption statistics are generated and passed to Monitoring & Prediction.
- Modify Production so that feedback is provided from Product Testing to Stock Control.

- Modify Planning so that production schedules are copied to Stock Control.
- Establish the changed relationship with each Supplier.

◇ *Identify target areas for IT support. Look for areas that (1) involve large quantities of data, (2) are time-critical or (3) involve complex processing.*

- Monitoring & Prediction is an obvious target for IT support, since it satisfies all three of the above-mentioned criteria.
- Monitoring & Prediction's reliances on consumption statistics and product updates are candidates for support—frequencies and data quantities need to be considered.
- The movement of parts to production units is necessarily to do with the movement of physical items, so IT support is not relevant.
- The notification of deliveries does not satisfy any of the three above-mentioned criteria; also, recording deliveries needs to be done 'at the factory gate' (although hand-held computers might be used).
- Scheduling of orders depends on the level of integration between Monitoring & Prediction and Orders & Invoicing.

In this particular example, we take Monitoring & Prediction, and the consumption statistics, to be the target for IT support.

16.7 Behavioural requirements

We are now in a position to clarify the corresponding behavioural requirements of the Monitoring & Prediction role.

◇ *Model the information involved in the New World behaviour.*

The services of delivering ordered parts, scheduling of orders, and notification of delivery all involve orders for parts, of specified kinds and quantities, from suppliers (Figure 16.4).

The provision of product updates is shown in Figure 16.5. In this model fragment, we encounter *ProductSpecifications*. For a given kind of product, these define one stage in the manufacturing process and the kinds and quantities of parts and materials involved in that stage. A *ProductionRun* involves a *ProductionUnit* taking products at a particular point of manufacture, and performing the next stage.

Production schedules are issued by Planning. These detail the production run for each Production Unit for the specified period (Figure 16.6). We can now see that the consumption statistics can be derived by Monitoring & Prediction from (a) the parts specifications within the product specifications that are referenced by production schedules, and (b) the number of products actually processed in each production run.

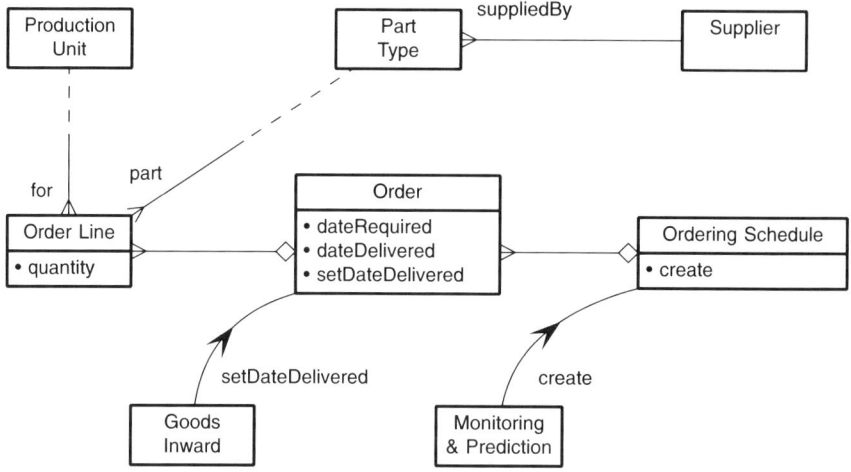

Figure 16.4 Ordering parts

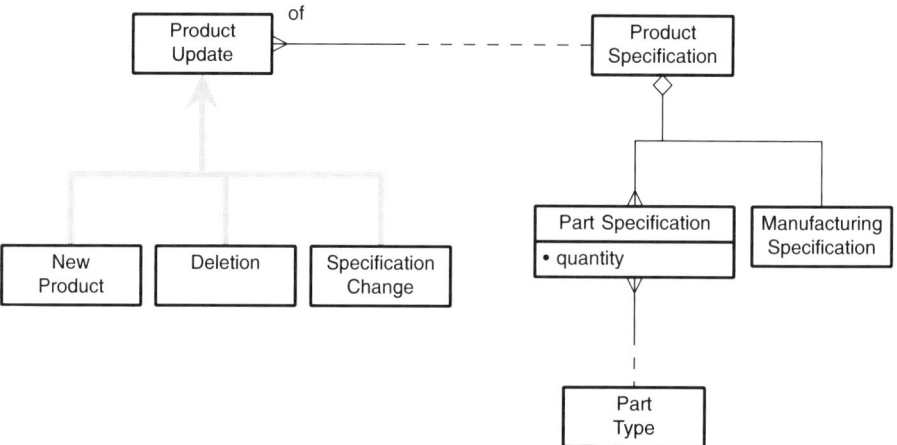

Figure 16.5 Product updates

Since the Production Units guarantee to follow the production schedules, the former information can be provided directly by Planning to Monitoring & Prediction. The Production Units simply need to record the number of products processed in each production run, and pass this information to Monitoring & Prediction.

◇ Determine the procedures (repeating behaviours) involved in the New World. Are they independent? When do they occur?

The main procedures are:

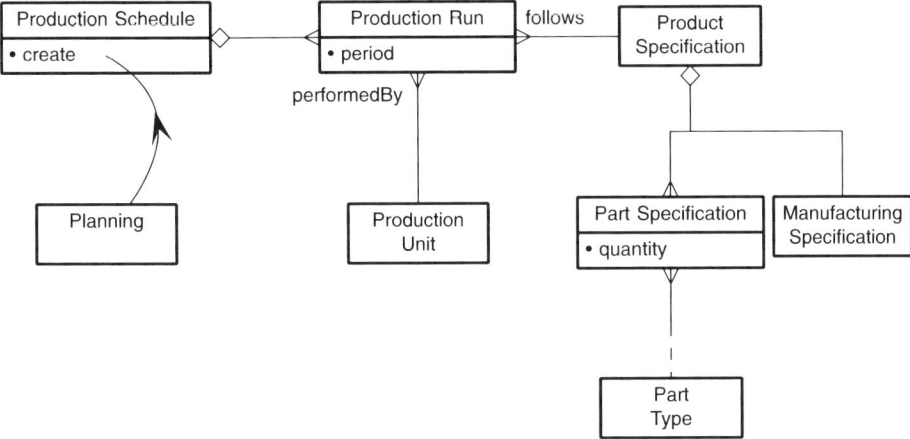

Figure 16.6 Production schedules

- collate production run data
- collate product update information
- generate ordering schedules

This first procedure is presumably tied into the production schedule periods; information will be supplied on a regular basis, as it is generated (for example, daily). Product updates are likely to be rare, so the procedure of collating this information can be done as and when updates are produced by Planning. The frequency of the third procedure, generate ordering schedules, depends on a number of quantitative factors: the rate of consumption of parts and materials vis-a-vis the minimum quantities and maximum frequencies of deliveries. A detailed analysis is required here.

There are also three supporting procedures:

- monitor Supplier quality performance
- monitor Supplier delivery performance
- maintain Supplier-part information

The first of these processes is activated by notification of faulty parts or materials by Production; presumably, some kind of fault report is generated and passed to the appropriate Supplier. The second process will probably be periodic (monthly, say), reviewing any differences between date-required and date-delivered attributes for completed orders, and flagging persistent offenders. The third process is triggered by changes of Supplier, changes to Supplier characteristics (for example, address), or changes to parts characteristics (for example, price).

◇ *Determine how interactions between the target IT system and its environment is to take place. What form of communications is used?*

The form of communications to be used for an interaction depends on factors such as the quantity of data and the frequency of the process consuming the data. For example, the communication of parts consumption statistics might benefit from an IT based solution because of the large quantities of data, the high frequency (daily or more frequent) and the physical distribution of the Production Units.

16.8 Development and transition

The behavioural requirements outlined above provide the basis for an overall design for Monitoring & Prediction. The usual design issues need to be considered: data storage and access, user interfaces, communications, and so on. In addition, there are several other issues that need to be addressed.

◇ *How does the target IT system integrate with existing software (where this is to be retained).*

In this case, the Old World Orders & Invoicing created orders, and tracked delivery, invoicing and payment. Is this worth keeping? Or is it so closely tied to the functions of Monitoring & Prediction that a unified system is preferable? If the IT components are to be separate in the New World, how is information passed between them?

◇ *Identify target users. Who are they? Will training be required?*

Will the users of the Monitoring & Prediction IT system be the existing Orders & Invoicing staff? Will they need to know more than they currently do? Will they need training in order to handle the chosen style of user interaction?

◇ *Produce a plan for transition from the Old World to the New World. How is the target IT system to be installed and brought on-line?*

Initially, Monitoring & Prediction will have no consumption data from which to generate ordering schedules. But, clearly, the Old World organisation cannot be dispensed with until the generation of ordering schedules is fully operational. One possible solution is to run Monitoring & Prediction in parallel with the Old World systems, but with only the consumption statistics and product update services operational. That is, Orders & Invoicing continue to generate orders on the basis of stock level reports from Warehousing. When sufficient data has been accumulated, Monitoring & Prediction could start to generate ordering schedules. Since the Production Units would then start to receive parts and materials without making explicit parts requests, Warehousing could be run down. Existing stocks could be used up by treating Warehousing as a temporary 'pseudo-supplier', receiving orders and delivering items from stock.

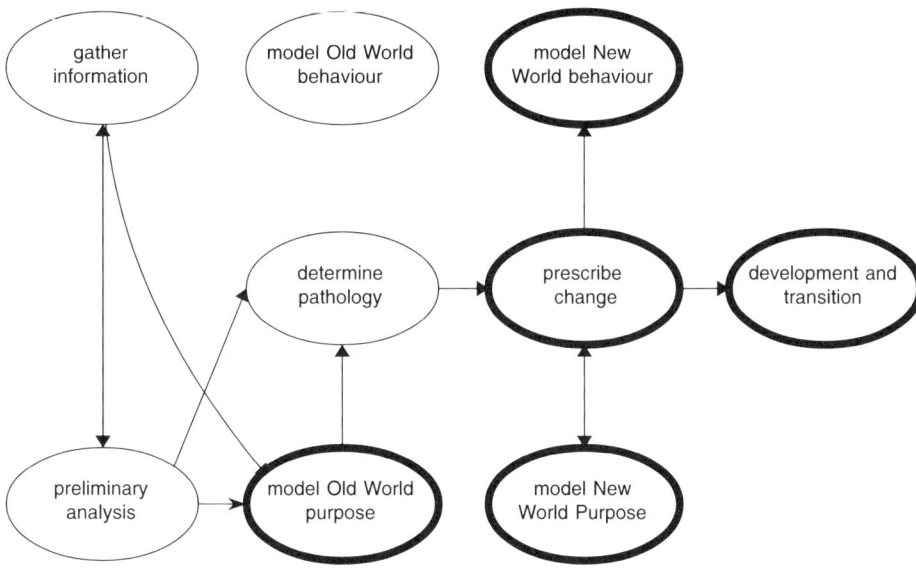

Figure 16.7 The tailored process

It is also worth noticing that the quality-related changes are independent of the delivery-related changes. Testing can be removed as soon as the changed relationships with Suppliers are established and Product Testing has amended its procedures.

Chapter 17

A new Purpose in life— the Paperless Map

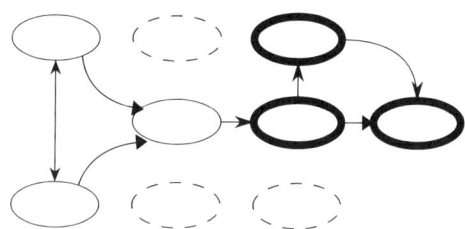

17.1 Analysis parameters

Typical scenario

- The analysis domain is an established organisation that is producing well-defined products or services.
- The environment has changed, and the organisation needs to change its purpose to 'keep up with the times' (to exploit the opportunity of new co-operations), or has had a new purpose thrust upon it (because existing co-operations have become problematic).
- The client does not want a fundamental change to the nature of the business.
- The business is based around some large, valuable, data-rich resource that currently makes little use of IT support.

Aims of the analysis

- The overall aim of the analysis is to identify how the valuable data resource could be better exploited to support the organisation's change of purpose.
- Specifically, the analyst needs to build an abstract model of the data resource that supports the new purpose as well as the old.
- The analyst should propose recommendations for the data migration task and the IT facilities required.

17.2 Preliminary analysis

◇ *Determine the reasons for analysis*

Our client is Rumaco (the Rumbabwe Mapping Company). In the past, they have had a near-monopoly on providing (paper-based) maps of Rumbabwe because of their large investment in country-wide survey data. The high entry cost of obtaining comparable data made it difficult for competitors to provide an alternative supply of maps. Now, however, technology has advanced, and map users are starting to agitate for electronic maps. Rumaco perceives a threat from start-up companies using computer-based Geographical Information Systems (GIS).

Rumaco want to utilise their current survey data by converting it to an information-base for use in their own GIS. This will allow them to meet customers' demand for electronic maps and other new products. This data re-engineering will also lead to a better exploitation of their own significant assets, hence again increasing the entry cost to competitors.

◇ *Determine the scope of analysis (what is and isn't open to consideration).*

Rumaco have determined their problem, and want the analysis to propose a solution based on computerising their existing survey data.

◇ *Determine the scope for change.*

Rumaco want to stay in map-making. New map-related products are necessary, but a radical change in the business direction is not permissible.

◇ *Arrange information gathering.*

The primary sources of information are twofold: those familiar with the current structure of the survey data, and those who are envisioning new products that will use the re-engineered information base.

17.2.1 Describing the Old World

◇ *Use purposive modelling to understand co-operations*

Rumaco is a traditional manufacturing business, with a co-operation with its Customer Base to supply and buy its products. This co-operation has become problematic at both ends: it is not supplying what its customers now want (electronic maps), and its customers are not buying what it does supply (paper maps).

The problematic co-operation is well-understood, so no further purposive modelling is required and we refrain from drawing the obvious Grampus diagram.

◇ *Use behavioural modelling to understand the current operation of the business*

The Old World survey data is mainly on paper charts. Some has been digitised, but in an unstructured and low level form: each map has many lines on it with feature codes to indicate whether it is a road, a building edge, a power line, ..., but there is no attempt to link these lines into larger objects, or to recognise true real-world things. Also, a single line might represent the edge of many objects, such as a road beside a river, but might be encoded as only one of these.

This lack of structure makes it infeasible to produce electronic products that differ significantly in functionality from traditional paper-based maps.

17.2.2 Determining the pathology, prescription for change

◇ *Identify the properties of the Old World that are blocking the desired changes.*

The current format of the survey data is too low level. An abstract information model is needed that can be used to structure the data in a more intelligent manner, in order to allow it to be used to produce new kinds of map-based products.

17.3 New World behaviour

The existing survey data must be re-engineered to derive knowledge about the real-world things that have been mapped. The first stage of this process is to construct a Beluga model of the information required in the New World.

The mapped things in the real world are Geographical Territories. These can be modelled as aggregations of Geographical Things like House, Road, City, River, etc. There are many hundreds of relevant subclasses of Geographical Thing. Geographical Things are abstracted as some geometric shape. These shapes are represented on a map, using rules about the line thicknesses, colours and shadings to be used for which shapes, and rules about what to do if shapes are too close or overlap.

For the purposes of map making, a Geographical Thing can be abstracted in a number of ways. For example, a road called 'Highway 42' can be abstracted as a line at a small scale, as a number of lines and nodes at a larger scale, and as an area consisting of a set of polygons at an even larger scale. In a similar way, a city such as Rumbare can be abstracted as a point or as an area. So, too, a house can be abstracted as a point or an area. Finally, any Geographical Thing can have a name abstraction, such as 'Highway 42' or '17, Alfalfa Avenue'. So the relevant subclasses of Geographical Thing Abstraction include: Volume, Area, Line, Point, Name and Image (a photograph is an example of an Image.) The Geographical Thing Abstractions are aggregated into an Abstract Map that corresponds to the Geographical Territory.

Each Geographical Thing Abstraction has a representation that depends on its abstraction. For example, a volume abstraction is represented as one or more solids, an area abstraction as a series of polygons, a line abstraction as a series of links, a point abstraction as a node, a name abstraction as a text string, and an image abstraction as pixels. So the relevant subclasses of Geographical Thing

17.4 Development strategy

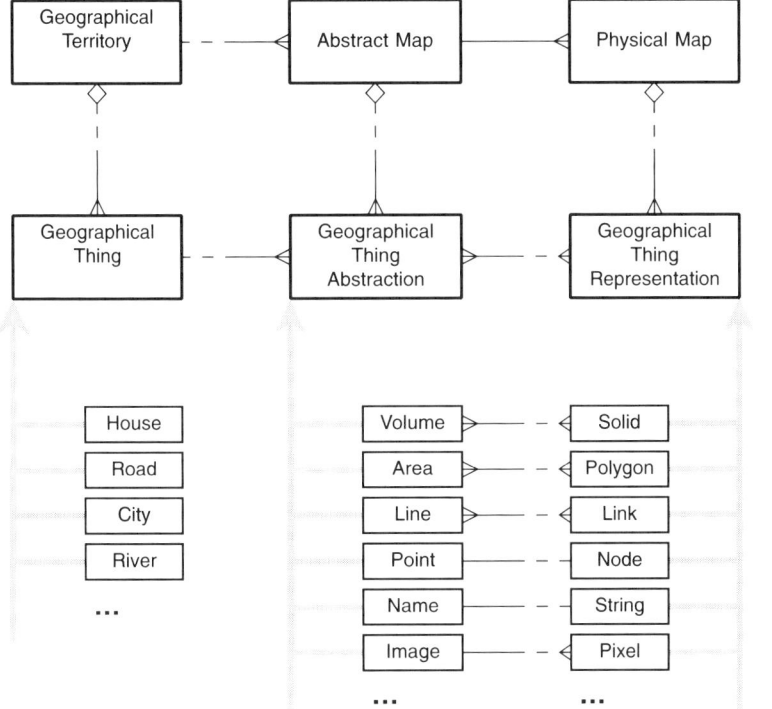

Figure 17.1 A geographical framework

Representation include: Solid, Polygon, Link, Node, String and Pixel. The Geographical Thing Representations are aggregated into a Physical Map (piece of paper, image on a screen, ...) that corresponds to the Abstract Map.

A Beluga model of this is shown in Figure 17.1. In the New World we no longer work at just the Representation level, but also at the more structured Abstraction level. New products can be designed using abstraction level data and concepts. The model can be further specialised as Rumaco develops new products.

17.4 Development strategy

The Rumbabwe public's changed reliance has required a change of Rumaco's purpose. The New World map information structures have to change to provide new products to meet the new customer requirements. The development strategy is:

- Re-engineer the Old World behaviour (consisting of just static information) into New World behaviour, by raising its level of abstraction from Representation Thing to Abstract Thing. This activity will include a large data

186 Chapter 17 A new Purpose in life—the Paperless Map

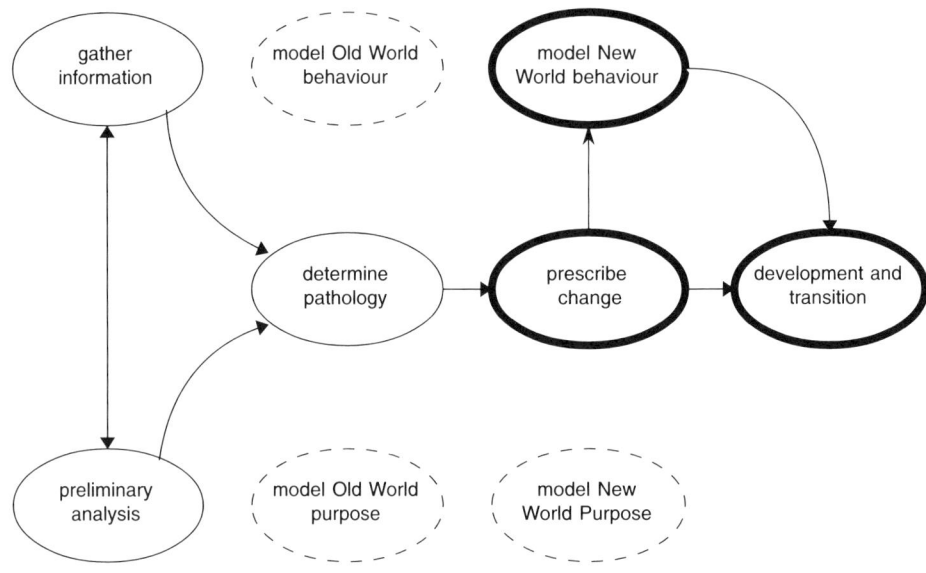

Figure 17.2 The tailored process

capture effort, to transform the current survey data into the new format.
- Use the higher level of abstraction to add dynamic behaviour as appropriate (for example, 'intelligent' maps could include a history of a house from initial planning, through construction, occupation, modification, to demolition).
- Generate the New World purpose, which is to sell new products (new kinds of Physical Maps) and the raw information (new kinds of Abstract Maps).

The tailored process is shown in Figure 17.2.

17.5 Enhancing our kitbag

Having successfully completed our analysis for Rumaco, we return to our office, and reflect on the models we have developed. We recognise that the geographical model is actually a specialisation of a more abstract framework (Figure 17.3). The classes in the top layer of the diagram correspond to the real-world model (giving the *semantics* of the model), its abstractions (*abstract syntaxes*) and its representations (*concrete syntaxes*). Each is an aggregation of the corresponding primitives.

We add this generic model to our analysis kitbag, where we can use it as a basis for the development of a model for a different problem area. (See Chapter 19 for a

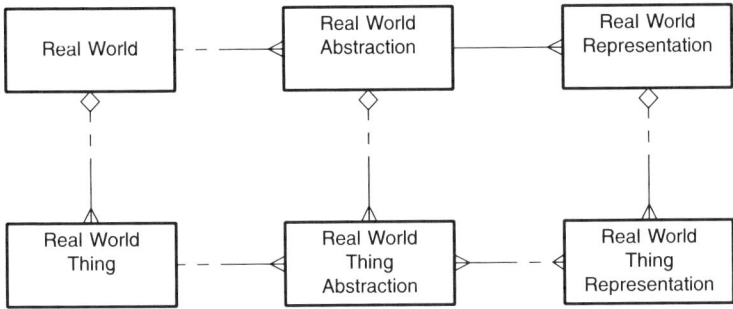

Figure 17.3 A more abstract framework

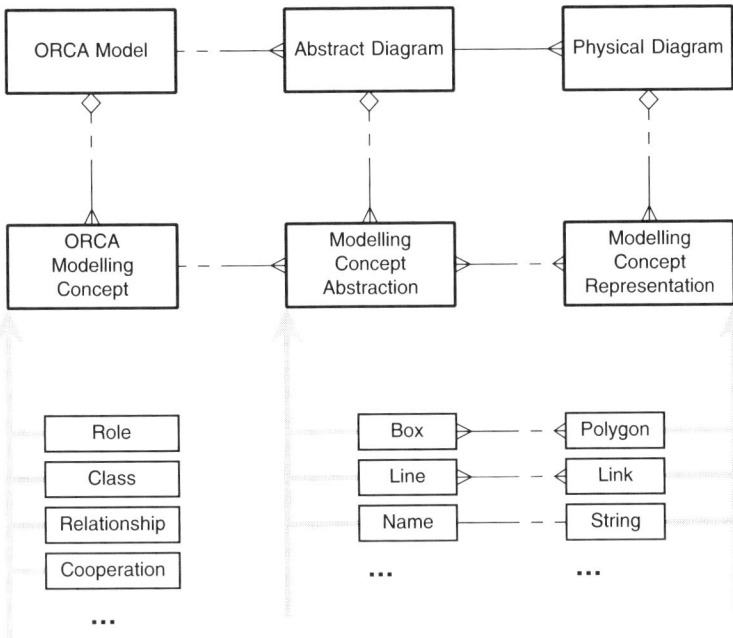

Figure 17.4 The framework specialised for ORCA

discussion of this approach.) For example, consider the design for tool support for ORCA. Here the 'real world' consists of Beluga and Grampus models, which are aggregations of classes, relationships, co-operations, and so on. The abstraction layer is of abstract diagrams (with components that are boxes, lines, and so on), and the representation layer is the actual concrete diagrams or textual equivalents. The framework specialised for ORCA is shown in Figure 17.4.

Chapter 18

No Old World—spreadsheets and telephony

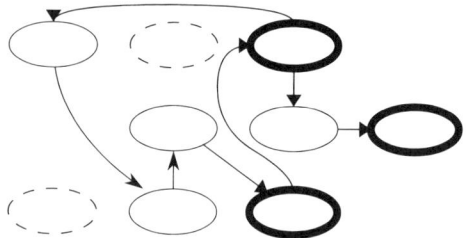

18.1 Analysis parameters

We have been using the term *Old World* to refer to the system that exists before analysis, and consequent change and development. However, there are situations where there is, apparently, no Old World (or, if there is, it seems irrelevant). It is still possible to use the ORCA approach in such situations, but the nature of the analysis is somewhat different from the Basic Process. In particular, the relationship of analysis and design is different. This chapter examines the issues raised by 'No Old World' analyses.

18.2 Why omit the Old World?

Why might we want to leave out consideration of the Old World? On one hand, we might have a situation where there is no Old World of any significance. On the other hand, the Old World might clearly exist but seem of no relevance to the new situation. That is, determining a system pathology and specifying the way forward would not be of any particular use. We now look at these alternatives in more detail.

18.2.1 The case of the missing Old World

Complete absence of an Old World is extremely rare. This occurs only when the design of a genuinely new product is being contemplated. Such brand new developments occur very rarely, especially in the more established areas of human activity (a status which Information Technology is gradually attaining).

As an example, consider the situation in the early years of the twentieth century in the heyday of the early motor car. In the 20 or so years that followed the appearance of barely acceptable vehicles, a vast number of completely new ideas appeared in the field. There were front-wheel, rear-wheel and all-wheel drive cars. There were cars with three or four wheels, cars with the wheels in a rectangular pattern, and cars with the wheels arranged in a diamond shape. Internal and external combustion engines were tried, with just about every possible combination of number and configuration of cylinders, induction and exhaust systems.

Slowly this free-for-all calmed down, and we ended up with the standard configurations that we see today. Car manufacturers still put great sums into research and development, but their efforts usually centre around the improvement of existing ideas rather than completely new developments that have no Old World to reference.

Software development is now in an era where the early flood of new ideas has diminished. The battleground for product and system suppliers is now concerned with developing effective products and well-integrated systems. Completely new ideas seldom occur.

Thus, situations where there is really no significant Old World are historically very rare. More common is the case where an opportunity is perceived for development of a product that fits into a 'hole' in the current world. In this situation, we should be able to model the environment that will provide a context for the new product, but not an Old World as such.

In particular, it can be useful to construct a purposive model of the new product's intended 'users'. What is it that users will require or expect of the product? Why will they want to use its services? In addition, it is important to consider the social context of the new development. How do the services provided by the product fit in with what users currently do? Will users be prepared to change the way they do things?

18.2.2 The irrelevant Old World

A different scenario is where there is an obvious Old World, but this turns out to be irrelevant, and possibly misleading, with regard to a new development. There is much evidence to suggest that a common problem in software development is the production of IT systems that closely match some existing, perhaps manual, system (the 'obvious Old World'), but fail to deliver the desired business benefits. This is because the developments impose no real change on the system. It is therefore crucial for an analyst to determine how much of the Old World is really essential to the system, and (if appropriate) make it clear that radical change is desirable.

This issue is closely tied up with that of Business Process Re-engineering (see Chapter 16). This process explicitly attempts to determine 'from first principles' how a business should operate, rather than blindly following existing structures and automating existing procedures.

As an example, consider the historical development of aircraft. The original attempts at designing flying machines leant heavily upon the 'obvious Old World'—in this case, birds. This produced a variety of bizarre contraptions that attempted to create lift and thrust by flapping their wings. Eventually it was realised that although wing-flapping was highly efficient if you were a bird, there were other ways of making progress if you could, for example, machine metals. This insight led to the development of modern fixed wing aircraft. The mistake that was initially made was to attempt a direct replacement for the 'obvious Old World' of birds. What *was* relevant was the less obvious, and more abstract, Old World of physics and aerodynamics, to which both birds and aircraft are subject.

We can now see that the real problem is not the irrelevance of the Old World, but the difficulty of deciding which bits of the Old World are truly relevant to the analysis task.

18.3 Modelling or design?

Broadly speaking, the ORCA method advocates that analysis precedes development and is distinct from it. The aim of analysis is to provide a context, a rationale and a specification for development. However, there are situations in which it is preferable to treat analysis and requirements definition as more nearly concurrent with design—when analysis of the Old World would not determine the development requirements to any useful degree. In this chapter, we have already discussed two scenarios in which this is the case: novel product development and radical business change.

Where IT requirements are largely under the control of the developer, design activity has the ability to promote understanding of a problem, make ideas more concrete, and raise or eliminate possibilities. For at least some initial part of the development process, there can be interaction between the analysis and design activities.

Support for this view comes from studies that have observed the progress of small development teams tackling IT problems. In one particular study, the rationalised view of the development process consisted of a succession of phases: Requirements Definition, High-Level Solution, Medium-Level Solution and Low-Level Solution. Requirements Definition concerned the 'application domain', while the Solution phases concerned successively more concrete levels in the 'computation domain'. However, the observed locus of activity did not follow this stepwise descent through levels of abstraction. Instead, there were frequent and dramatic leaps between levels. In particular, insights about the application domain and requirements were frequently preceded or followed by bouts of very concrete design and experimentation. Activity at the higher design levels was relatively rare until the later stages of development.

In conclusion, then, we can say that where circumstances do not allow analysis to precede development, the two activities can be interleaved. However, this should be the exception rather than the rule, since it complicates project organisation and planning. The study described above dealt with a small-scale development exercise, not a large-scale development project. Chapter 21 discusses the related topic of prototyping, together with other project-related issues.

18.4 No Old World processes

In this section we describe two 'No Old World' processes that are applicable in the sorts of circumstances discussed in this chapter.

18.4.1 New product—the spreadsheet

The personal computer (PC) spreadsheet is now an everyday part of computing—so much so that it is difficult to remember the fuss when such products were first released. The spreadsheet was one of the first really new ideas that were enabled by the existence of the PC, and it has been credited with a good deal of responsibility for the current prevalence of these computers. (Previous spreadsheet-like packages did exist, but they lacked the immediacy of manipulation that characterises the modern implementation.)

It is instructive to consider the application of ORCA to the development of a spreadsheet. According to interviews with the original developers, the original notion was of some sort of 'super calculator'. Design proceeded using this notion until thoughts moved on to possible applications of the product. It was then that the Old World (in ORCA terms) of balance sheets and accountancy ledgers was assimilated, and we ended up with the first versions of the products that we have today.

In summary, the process that was followed in this situation proceeded in the following manner (Figure 18.1):

1. Design the essential elements of the new product, in order to understand what is feasible and some of the capabilities of that device.
2. Abstract up towards the requirements domain.
3. Define the parts of the Old World relevant to the new product.
4. Define the New World, and specify and design the new product.

18.4.2 Radical change—the introduction of telephony

According to popular legend [Gardner 1990, Chapter 25] the Chief Engineer of the British Post Office greeted the invention of the telephone with the observation:

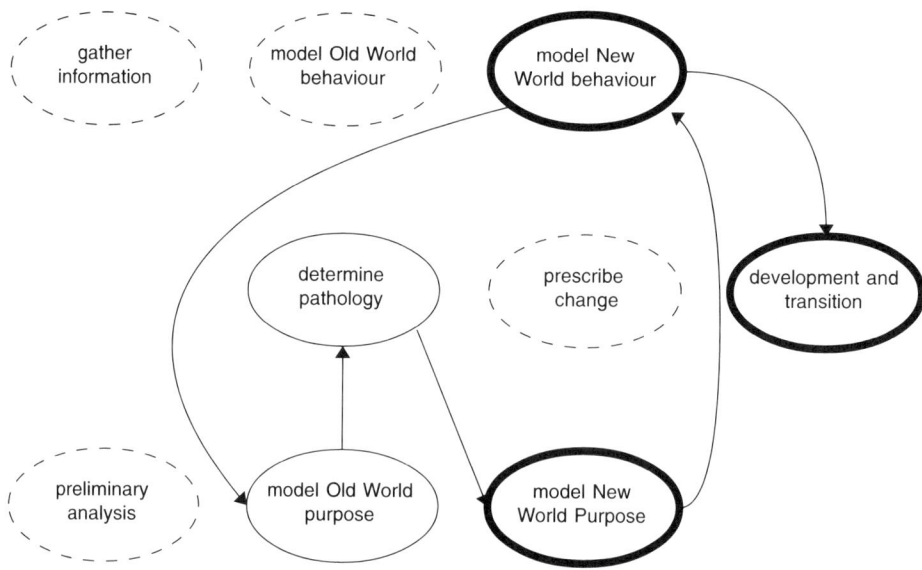

Figure 18.1 A process tailored for a new product

"The Americans have need of the telephone—but we do not. We have plenty of messenger boys."

This is a classic case of the imposition of an irrelevant Old World. Here the mental image of business communication was strongly influenced by the existing communication media, to the extent that the functionality of a new medium was stated by comparison with those existing approaches.

Since the prevailing model of communication was one in which an underling was dispatched with a particular message to a particular destination, with no hope of a continuing conversation, then the impact of a circuit-switched telephony conversation would be difficult to imagine. (Interestingly, the introduction of packet switching to computer networks late in the twentieth century, analogous to going back to a messenger style of communication, has many advantages for computer communication.)

In this situation we can sketch what would have been a more suitable approach to the introduction of telephony (Figure 18.2).

18.5 Summary

The rest of this book implicitly assumes that the process of requirements capture and analysis is strongly driven by modelling some of the Old World. However, it is

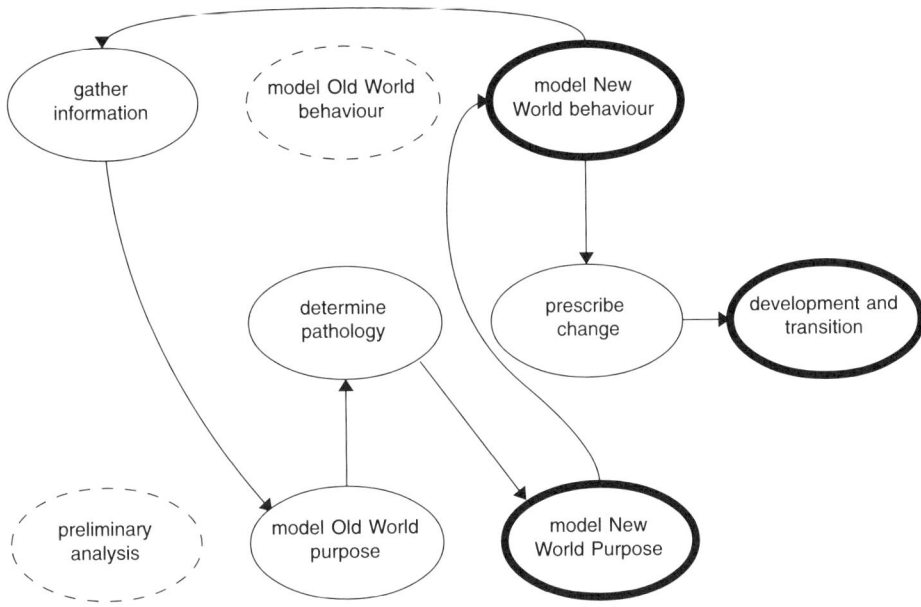

Figure 18.2 A process tailored for a radical change

sometimes difficult to determine which parts of an existing situation are genuinely part of an Old World, and which are excessive detail that might well obscure a useful result. Much has been written about the apparent failure of IT to deliver the promised improvements in cost-effectiveness. Many of the situations where this is the case have applied a comprehensive model of the Old World blindly to a new development—most obviously in the simplistic automation of existing business processes. In these situations the development of a model, at a suitable level of abstraction, of what was important in a particular situation could well have produced more useful results.

In situations where requirements are not determined to a useful degree by analysis of the Old World, interleaving analysis and design activities can be a useful way of gaining insight into a problem and establishing the technological capabilities available to a developer, determining what really is the Old World.

Chapter 19

Nothing new under the sun— a lending library

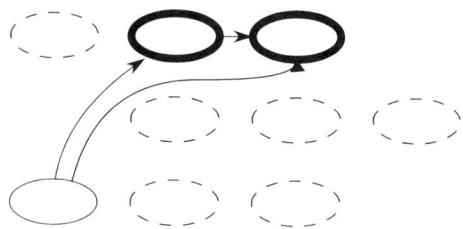

19.1 The analysis strategy

This chapter explores a strategy for behavioural modelling that is potentially applicable in a variety of analysis situations. The strategy is to adapt existing generic models to reflect the actual systems being analysed.

It is fairly obvious that there are commonly occurring categories of system, and that systems within the same category have similarities. For example, we often refer to 'accounting systems', 'stock control systems', 'transaction processing systems', 'process control systems', and so on. It follows that we can, in principle, have generic models for categories of system. In order to do this, we need a modelling language that allows sufficiently abstract models to be constructed.

We can envisage an analyst using a 'kit bag' of generic models to drive behavioural modelling. The effectiveness of this strategy will depend on the analyst's expertise in recognising situations where generic models are applicable. An analyst will need to have considerable experience and understanding of the kit bag contents. Of course, an analyst does not have to get it right first time; different candidate models can be tried out as working hypotheses. Candidate models can be used to prompt analysis questions as part of the process of adapting the model for a particular situation. Models can be used to check coverage—if a generic model has a component that appears to be missing in the actual system, is it genuinely absent, or has it just been overlooked?

Adapting a generic model to reflect an actual system might involve extending the model, restricting it, specialising its components, or combining it with other model fragments. The process of adapting generic models should be as rigorous as possible, otherwise the connection with the original model is weakened or lost. This would call into question the analyst's judgement that the generic model was applicable in the first place. A strategy of adapting generic models is not an all-or-nothing approach; we might well want to build up a complex model using both suitably adapted generic models and bespoke models invented for the particular analysis. For example, an aggregation often consists of a bespoke aggregator class that adds highly specialised behaviour over and above that provided by more generic aggregated classes.

The next section illustrates these ideas using a simple example. Section 19.3 describes briefly some other generic models, and summarises the tailored process.

19.2 A lending library

19.2.1 Brief description of the domain

The domain for our modelling example is a Lending Library (a little more realistic than that in Chapter 12). For the purposes of this chapter, the aim of the analysis is to model and understand the behaviour of the Library. Ultimately, there may be a requirement to provide IT support to the librarians, but we choose not to consider this aspect of the process here.

Our Lending Library allows books to be lent to borrowers. There are restrictions on which books can be borrowed, on the duration of loans, and on the number of books held by a borrower at any given time. In addition, reservations can be made by borrowers against book titles (since there may be multiple copies of a title). A reservation guarantees a borrower's option to borrow a book at such time as a copy of the desired title becomes free. All this should be sufficiently familiar not to require more detailed description.

19.2.2 A candidate generic model

Ignoring the aspect of behaviour concerning reservations for the time being, the Lending Library behaviour seems essentially to be that of an ItemAllocator framework. The behaviour consists of multiple *Transactions*. In each *Transaction*, some *User* ($\longrightarrow$ borrower) makes a request to an *Allocator* ($\longrightarrow$ librarian) for an *Item* ($\longrightarrow$ book copy) of a specified *Type* ($\longrightarrow$ book title); if various constraints are satisfied, the *User* and a suitable *Item* are associated for some period ($\longrightarrow$ book on loan to borrower); this period is terminated by the *User* ($\longrightarrow$ borrower returns book). An *Item* is either *notInUse*, or is associated with a single *User*.

As luck would have it, we have a model of an *ItemAllocator* in our kit bag of generic models, which we produce with a suitable flourish ('and here's one I prepared earlier...'). At the class level, our *ItemAllocator* framework looks like Figure 19.1. The constituent sets are:

- *users* : $User^+$ // one or more, typically many
- *items* : $Item^+$ // one or more, typically many
- *allocator* : *Allocator*   // just the one

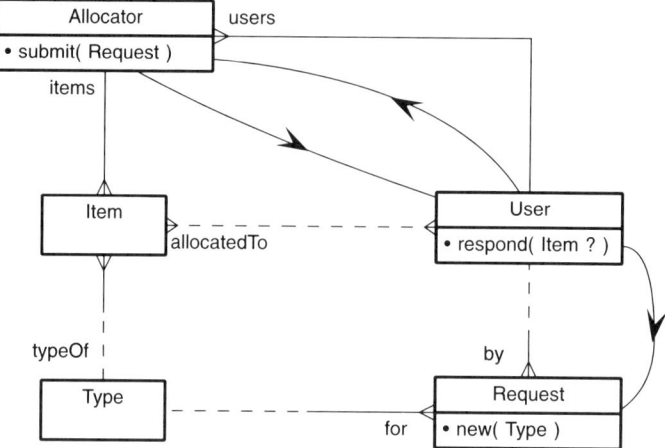

Figure 19.1 Item Allocator framework

- requests : Request* // zero or more

Note the following points. The cardinality of the *allocatedTo* association is many-to-many—the generic model does not prohibit the sharing of *Items*. *Requests* may contain information other than the desired *Type* of *Item*, for example a *Request* might specify a period for which an *Item* is required.

The dynamics consists of a composition of *Transactions*; the composition is parallel because *Transactions* can overlap. The structure of a *Transaction* is shown in Figure 19.2.

First, a user u creates a request r, which becomes associated with u and the desired type t of item i. Then u submits r to the allocator a, which responds to u, setting up an association between u and i (written as u—i) if the parameters of r are valid and a suitable i is available. If a u—i association is set up, it is subsequently dissolved; the duration of the association is labelled d.

The precondition for setting up a u—i association is shown as $valid(u, r, t, i)$. In fact, it consists of several subconditions:

- $cleared(u, t)$: is user u currently cleared to use items of type t?
- $suitable(i, t)$: is item i of type t?
- $available(i, r)$: is i currently available to satisfy request r?

As instigator of the *respond* interaction, the allocator a is responsible for ensuring the precondition.

Note that no interaction is specified with the u–x–i dissociation *release*. This deliberate vagueness allows three possibilities:

- u might relinquish i
- a third party (presumably a) dissolves the association

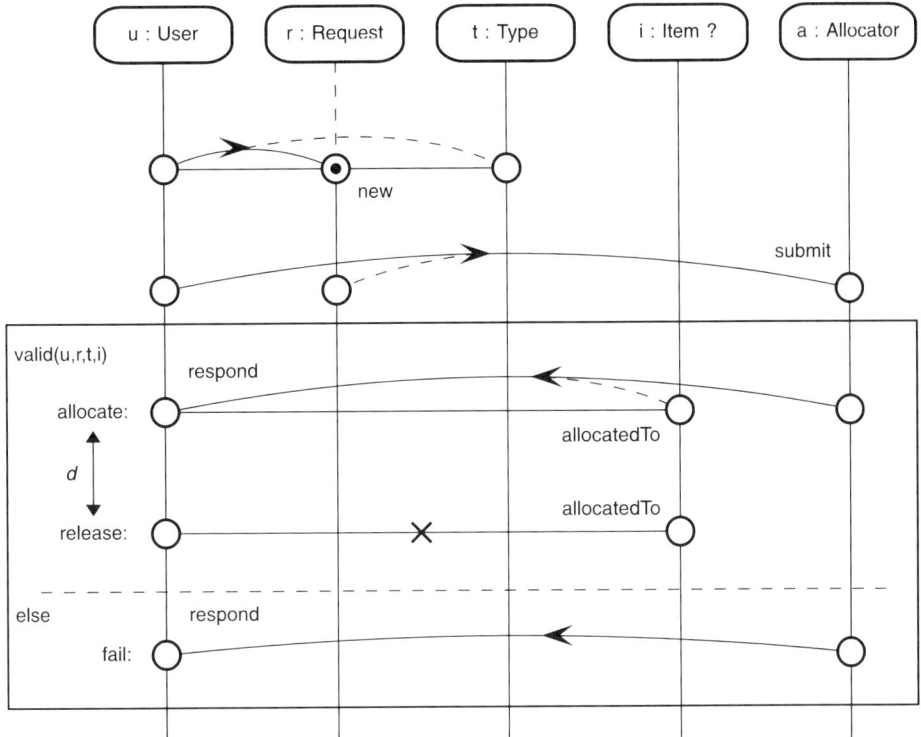

Figure 19.2 A Transaction

- *i* breaks the association (for example, a software loan deleting itself after a timeout)

19.2.3 Adapting the model

Renaming

The first and simplest stage of adaptation is to rename components of the model (classes, associations, events). For example:
- *Item* ⟶ *Copy*
- *Type* ⟶ *Title*
- *User* ⟶ *Borrower*
- *Allocator* ⟶ *Librarian*
- *Request* ⟶ *Request*
- *allocatedTo* ⟶ *onLoanTo*
- *typeOf* ⟶ *titleOf*

- *allocate* $\longrightarrow$ *borrow*
- *release* $\longrightarrow$ *return*

The names of constituent sets also need to be changed appropriately.

Features and subclasses

Do the classes in the model need to define additional features? For example, *Borrower* needs a feature specifying the maximum number of copies that can be held at any one time, *loanLimit*; *Title* needs a feature specifying the maximum value for the *loanDuration*, d. These features could be fixed at object-creation time, or there could be operations to set (or reset) them during the lifetime of an object.

Do we need to define subclasses, in order to reflect fixed distinctions? In our proposed class renaming (see above), *Type* was renamed to *Title*. However, in the Library, some titles are for reference books, which cannot be borrowed. So we need two classes: a class *Title* which has the *titleOf* association with *Copy*, and a subclass *LoanableTitle* for titles which can be referenced by a loan *Request* (the *RequestforLoanableTitle* association).

Note that the subclass *LoanableTitle* adds behaviour to class *Title* (the ability to be requested). We do *not* have a loanable *Title* class with a subclass *ReferenceTitle* that removes the loanability. This would be incorrect because instances of *ReferenceTitle* could not in general be treated as instances of their superclass.

Looking at constraints

In the generic model, the cardinality of the *allocatedTo* association is many-to-many. In the case of the Library specialisation, it is physically impossible for copies to be shared by many users at one time. The primary reason for copies being unavailable for loan is that they are already on loan. In other words, each copy is either *in* the Library, or *out* on loan. The condition *available*(i, r) becomes

$$inLibrary(copy) \text{ and } loanable(copy)$$

Since requests are for immediate loan, the availability condition does not involve a request r. In order to check this condition, the librarian needs to record the in/out status of each copy.

A borrower can borrow a copy provided that the loan limit has not been reached; this does not depend on the title requested. Thus the condition *cleared*(u, t) becomes *belowLoanLimit*(*borrower*). In order to check this condition, the librarian needs to maintain a record of each borrower's current loans.

The condition *suitable*(i, t) remains the same (copy c is of title t).

In addition, it is the responsibility of the librarian to enforce the maximum duration for a loan (given the limit specified for the title).

Elaborating the behaviour

In the generic model, it is not specified what happens to requests after their submission to the allocator/librarian. In the customised model, there are a number of cases that need to be considered:

- If the whole precondition is true, the request becomes a *loanRecord*; this would need to hold the start-time of the loan and the copy involved. The loan record is deleted when the copy is returned.
- If the borrower fails the *belowLoanLimit* test, the request is deleted.
- If a suitable copy is not available (all copies of the title are out on loan), the borrower has the option of making the request into a *reservation*. Reservations for a title are held in order of creation. The *return* behaviour needs to be elaborated to check for extant reservations; if there is an extant reservation for the title of the returned copy, the copy is held and the borrower who made the reservation is informed. The reservation is deleted when this borrower subsequently borrows the held copy. Alternatively, the reservation is deleted on expiry of a fixed holding period.
- If a request is made by a borrower who already holds a copy of the specified title, then it is treated as a request for renewal of the loan. The original request (now a *loanRecord*) is updated and the renewal request is deleted.

Allowing reservations modifies the availability condition. Copies now have three statuses: *free*, *held* and *out* (where $in = free\ or\ held$). A copy is available for loan only if it is *free*. A renewal is valid only if there are no extant reservations against the title.

Causation

The generic model does not specify who causes the *release* (return) behaviour. In the case of the Library, it is clearly the borrower who returns copies. However, this raises the issue of how the librarian enforces the maximum loan duration given that borrowers are free agents.

One possibility is for the librarian to notify the borrower involved in a loan on expiry of the maximum duration. This implies that the librarian carries out some periodic scanning of the loan records, looking for overdue loans.

Other issues

In the generic model, a *User* (borrower) associates a *Type* (title) with a newly created request. This implies that users have access to the set of types. How does this happen in the Library? In the normal case, a borrower takes a copy off the shelves and presents it to the librarian; the titles of copies that are on the shelves are thus implicitly accessible. But how does a borrower make a request for an unavailable title (that is, a reservation)? Some access to the full catalogue of titles is needed.

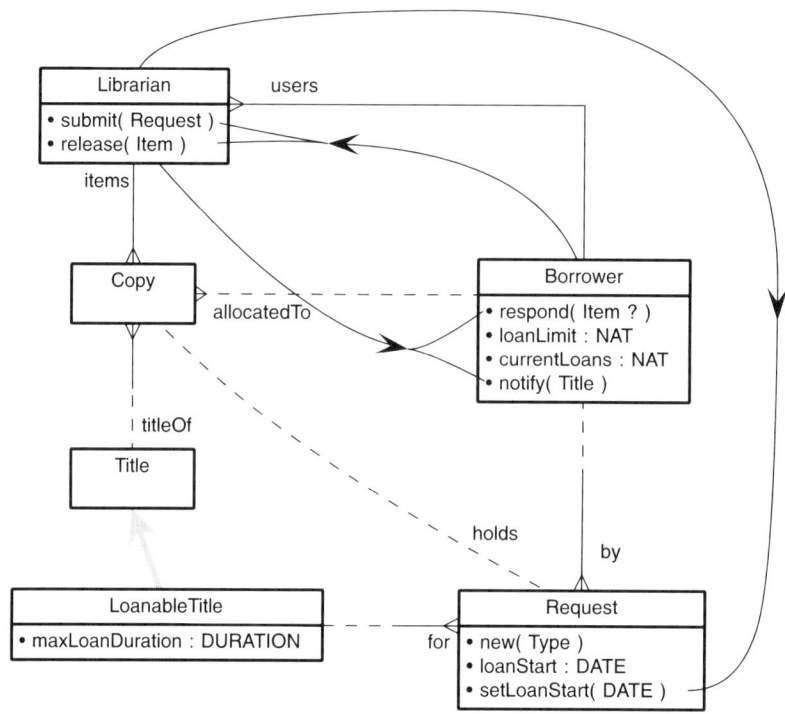

Figure 19.3 The Library framework

So far, we have been talking about 'the librarian', but our Library may well have multiple librarians. If so, how do they work together? It is important that the librarians have shared and controlled access to the records of titles, copies and borrowers.

The constituent sets for borrowers, titles and copies are not fixed—borrowers may be enrolled and disenrolled, titles may be added to the catalogue, copies may be acquired and withdrawn. There needs to be additional behaviours covering these aspects.

19.2.4 The adapted model

Our *Library* framework, in terms of its classes, is shown in Figure 19.3. Note the additional *holds* association between *Request* and *Copy*. The *allocatedTo* association can be derived from the *holds* and *by* associations. Note also the additional interactions.

The dynamics of the Library are a composition of the following:

19.2 A lending library

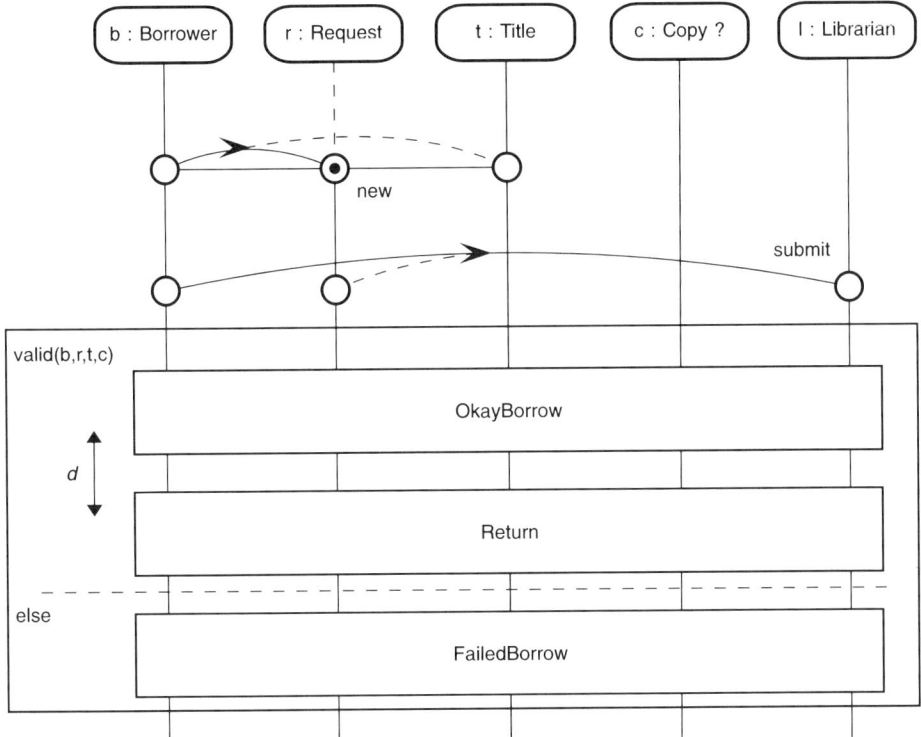

Figure 19.4 A Library Transaction

- multiple parallel loan transactions
- periodic scanning for overdue loans and notification of errant borrowers
- behaviour concerned with enrolling and disenrolling borrowers, adding titles to the catalogue, acquiring and withdrawing copies.

At the top-level, the structure of a transaction is the same as in the generic model, but some additional complexities have been factored out (Figure 19.4). The validity precondition can be expanded as follows:

$valid(b, r, t, c) =$
  $(free(c)$ or $held(t, c, b))$
  and $belowLoanLimit(b)$
  and $loanable(t)$
  and $t\ titleOf\ c$

where:

  $free(c) =$ not exists $(r_0 : Request)$ such that $(r_0\ holds\ c)$
  // no reservations or loans holding c

$$held(t, c, b) = \text{exists } (r_0 : Request) \text{ such that}$$
$$(r_0 \text{ holds } c) \text{ and } (r_0 \text{ for } t) \text{ and } (r_0 \text{ by } b)$$
$$\text{// c is held by b's reservation for t}$$

$$belowLoanLimit(b) = b.currentLoans < b.loanLimit$$

$$loanable(t) = (classOf\ t) = LoanableTitle$$

The subframework *OkayBorrow* is a composition of the following three behaviours:
- The librarian allocates the copy to the borrower.
- The librarian makes the request into a loan record by setting the *loanStart* and associating it with the copy.
- If the copy was held against a reservation (by b for t), then that original request is deleted, otherwise nothing.

These three subbehaviours are shown in Figure 19.5. There are similar subframeworks for *Return* and *FailedBorrow*, consisting of the basic behaviours (from the generic model), together with additional (possibly conditional) behaviours. The full Library model also needs frameworks for the other aspects of behaviour mentioned earlier.

This example shows how a suitable generic model can be adapted to model a particular system. The generic model is necessarily 'permissive' in various respects, since otherwise it could not be applicable to a range of actual systems. The essence of the process is to address these areas of uncertainty in the context of the actual system being described.

19.3 Other uses, other models

19.3.1 A different use of the Item Allocator model

A different use of the generic *ItemAllocator* model would be for a dynamic resource allocation system. Items would be resources, and Users might be tasks or processes. Some of the key differences are as follows:
- Resources can have multiple types (for example, staff with multiple skills in a project management system). Requests may specify compound types (for example, a person with skill A *and* skill B).
- Resources can be shared between tasks, where each task requires only partial utilisation of a resource. Requests will therefore specify the required level of utilisation.
- Requests can specify a duration for which a resource is required, and possibly constraints on start/finish times also.

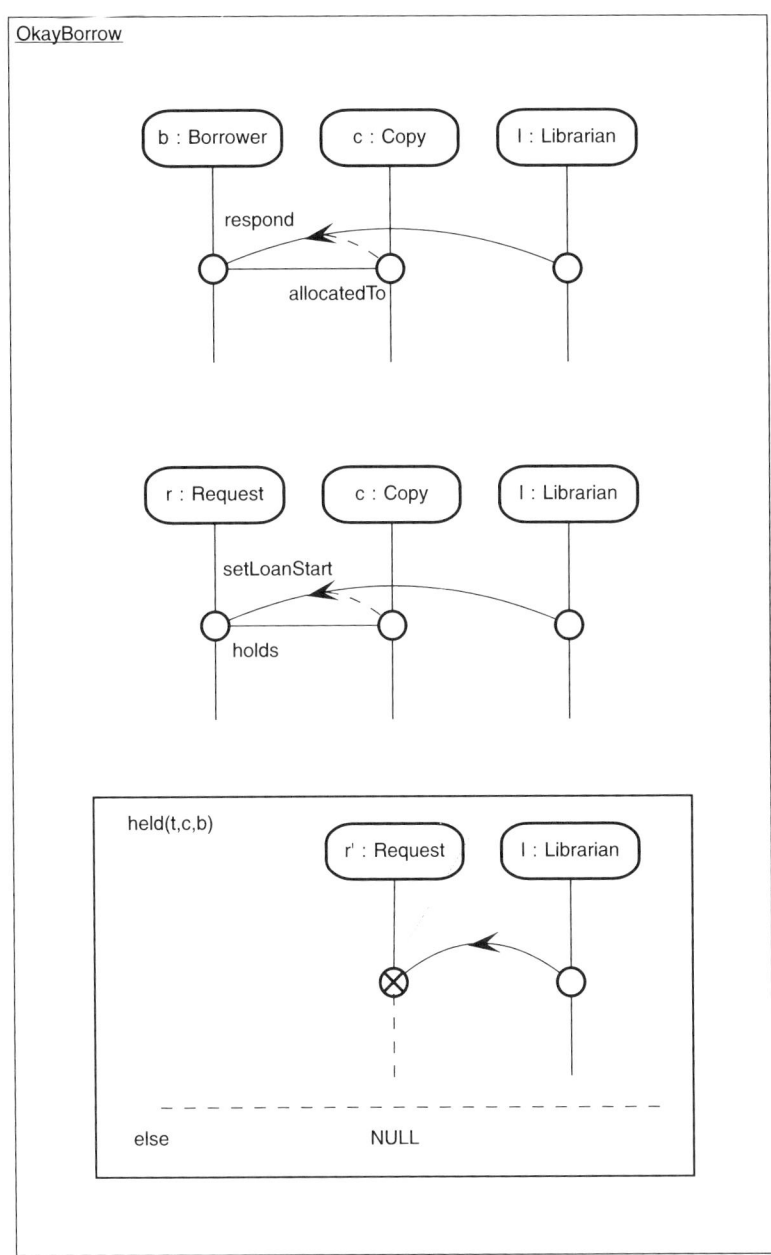

Figure 19.5 The subframework *OkayBorrow*

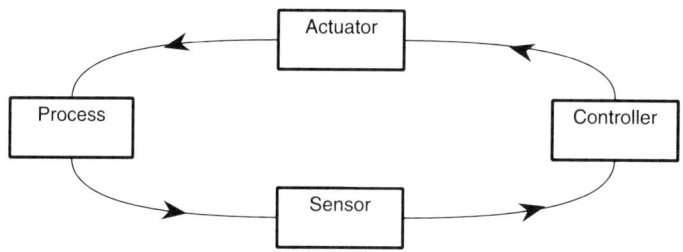

Figure 19.6 A Control System framework

The functionality of the allocator in this kind of system will be much more complex than in the case of the Library. In the Library, there is no need to select between alternative available items; if more than one copy of a title is available, it does not matter which one is selected. In a resource allocation system, this is not the case. If a request for a resource of type a is received, and resources of type $a + b$ and $a + c$ are available, then the choice of which one to allocate is significant. For example, if requests for resources of type b are much more common than requests for type c, then it makes sense to allocate the $a + c$ resource. These complexities are compounded if requests may be held pending release of resources from ongoing tasks.

19.3.2 Other generic models

The kinds of generic model that are suitable for an analysis kit bag need to trade-off abstractness against content. The more abstract a model, the more widely applicable it will be. However, the more abstract a model, the less it says about a system to which it is applied, giving less of a head-start to the analyst. And conversely.

This subsection outlines some candidate models that strike a balance between these competing requirements.

Control system

An example of a control system appears in Chapter 8, applied to the NIMWeC Looms. The class-level view is shown in Figure 19.6. An interesting feature of control systems is that the *Process* is essentially continuous; *Sensors* provides 'snap-shots' of the process to the *Controller* which decides on actions to be taken by the *Actuators*. As the diagram indicates, the chain of interactions is basically cyclical.

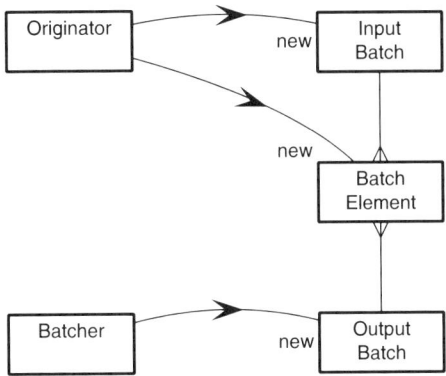

Figure 19.7 A Batching framework

Batching System

NIMWeC also involvs an example of a Batching System (Chapter 8). In generic form, input batches of elements are created by the originator; the batcher rearranges these into output batches, according to some criteria (Figure 19.7). For example, the elements of orders, generated by customers, may be rearranged into batches which can then by assigned to particular production units on the basis of the required product characteristics (model, colour, and so on).

Stock Control System

In a Stock Control System, consumers make requests to a store, specifying the type and quantity of item required; where possible, the store supplies these requests from current stocks; additional stock is obtained from suppliers (with some delay). Orders for replacement stock are made on the basis of past and predicted requests for items (Figure 19.8). The aims of the store are to minimise stock levels and to minimise the response-time to consumer requests (in other words, the number of requests that cannot be supplied from stock). These aims are clearly competing.

A variant of this occurred in the 'just in time' manufacturing system in Chapter 16.

Model-Views System

Many architectures involve behaviour that provides multiple views of an underlying model. In a Model-Views System, a succession of events impinges on the system, with each event being handled by a particular view (Figure 19.9). Different subclasses of event require handling in the following different ways:

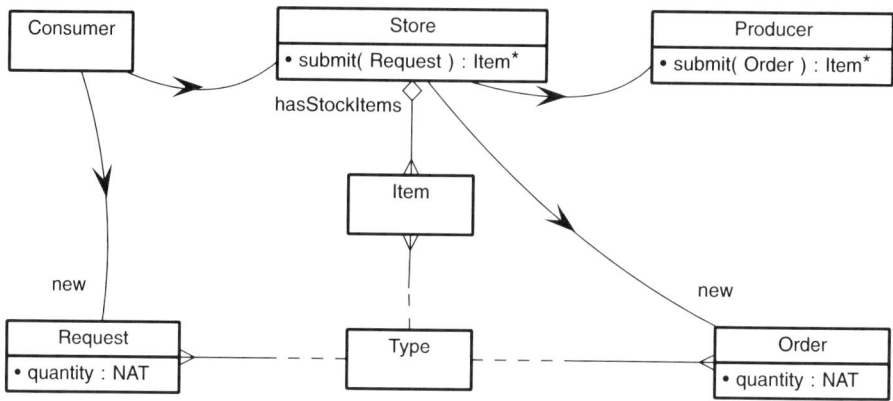

Figure 19.8 A Stock Control framework

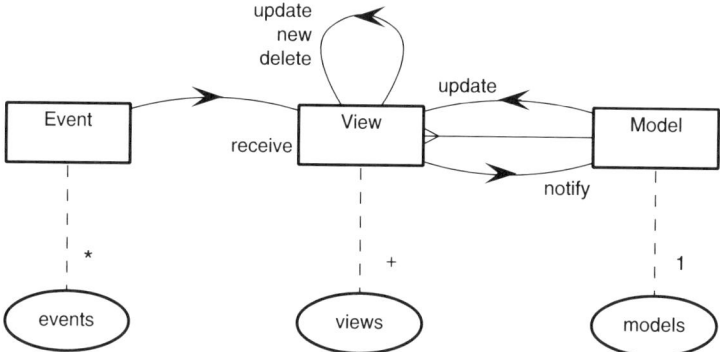

Figure 19.9 A Model-View framework

- Only the receiving view is affected.
- The event causes changes to the model, and propagation of these changes to all the views.
- The receiving view is deleted or a new view is created.

After handling each event, there should be consistency between views and model (that is, the views should be accurate, if partial, presentations of the model).

19.3.3 Building up an analysis kit bag

A useful kit bag of generic models will not come about by chance. One way to generate suitable models is simply to set about the task from scratch, attempting to provide generic models for known application domains. However, a one-to-

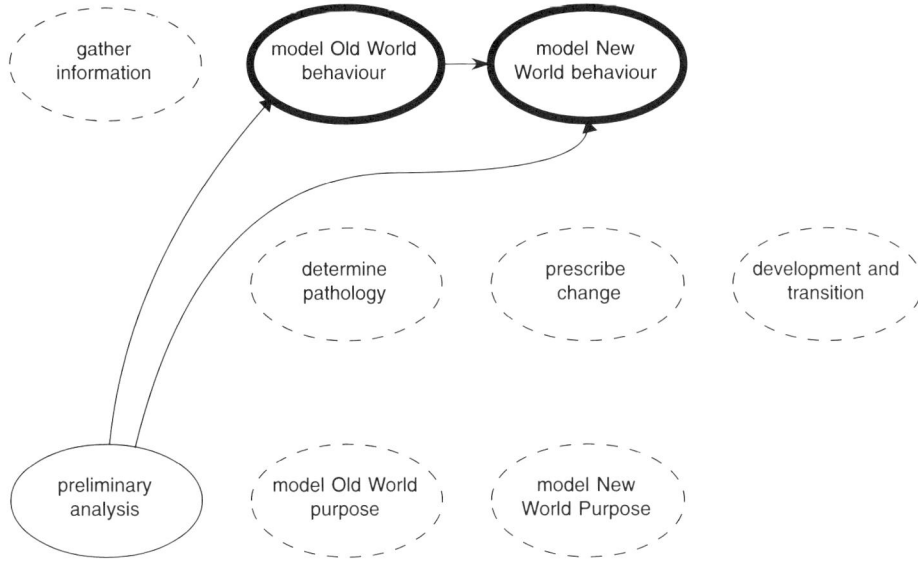

Figure 19.10 A process model tailored for reuse

one correspondence between application domains and models may miss underlying similarities. For example, it is not immediately obvious that a lending library and a resource scheduler share a common abstraction.

An alternative approach is to 'reverse engineer' bespoke models as and when they are produced (see Chapter 17, for example). This has the advantage that it is potentially a useful part of the analysis process in its own right. Part-way through constructing a bespoke model, we might realise that there are similarities with a model of another system, previously constructed. Useful insights might be gained by constructing a common generic model and using this to continue with the current analysis.

There is an analogy here with the reuse of software components—the more explicitly a component has been 'designed for reuse', the more reusable it will be.

Chapter 20

No-one to talk to—Ahab applied to NIMWeC

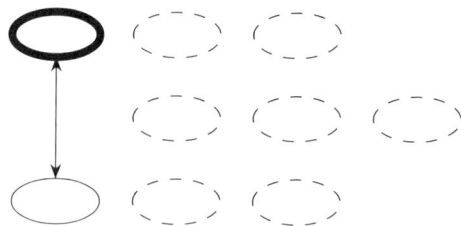

20.1 The need for tailoring

Many published OO techniques presuppose their starting point to be a correct and coherent textual requirements specification. In practice, this is seldom if ever the case. Analysts often need to understand a written account of some state of affairs, where the text is baroque, long, complex, incomplete or erroneous. Where the text is the only source of information—where there is no immediate access to the authors, to the client or to domain experts—then help is needed to analyse it.

In short, the problem is having to base the initial analysis on text which is *balderdash*. In such a case, ORCA's Basic Process is tailored by adding a technique, called Ahab (**A** **H**elp in **A**nalysing **B**alderdash), to the preliminary stages.

20.2 Balderdash

Confusing text can arise for several reasons.

Government-speak and jargon

Documents originating from government or defence sources often have an abundance of unfamiliar terms, strings of initial letters, acronyms, abbreviations and jargon. As the density of such obscure terms waxes, comprehension can wane. This is often accompanied by baroque style and eccentric grammar.

Synonyms, synecdoche and malapropisms

Multiple authors, or authors infected by 'creative writing' lessons, may use different terms to describe the same concept, leaving readers to wonder what subtle distinction is being made. They may use terms imprecisely, sometimes referring to one thing, sometimes to only a part of it, sometimes to a different but related thing. They may even use the wrong term. A consistent vocabulary needs to be established.

Length and complexity

Problem descriptions often need to be long and complex, and even a well-written text may for this reason be initially impenetrable. It can be difficult to keep all the important points clear, and mistakes and inconsistencies can easily be made and overlooked.

Incompleteness, and unnecessary detail

Authors commonly fail to define domain-specific terms and fail to state supposedly well-known assumptions. They assume that readers are more familiar with the domain than is the case. This is often accompanied by unnecessary detail about those topics that the authors are not so familiar with (often in the very areas of the readers' expertise), or irrelevant detail, to assuage guilt for earlier omissions.

20.3 Ahab

As well as neglecting the possibility of the text being composed of balderdash, many methods also place too much weight on the importance of nouns, and nouns alone, as the source for objects, classes or frameworks (a recognised problem in all forms of data modelling). Ahab addresses both these problems.

Ahab is based on simple linguistic analysis at the clause level. An Ahab diagram is used to represent a collection of statements and to group together related concepts. A simple statement, typically composed of two noun phrases linked by a verb phrase, is represented by an Ahab fragment. The noun phrases are shown in circles, linked by a line annotated with the verb phrase. An Ahab diagram is made up of a collection of these fragments, merged or overlapped at a noun phrase. Examples of Ahab fragments and diagrams are given in Section 20.4.

The main features of Ahab are as follows.

- It helps to separate the relevant parts of a text from the dross.
- It helps to identify ambiguities and inconsistencies, such as synonyms (words that mean the same thing but look different) and homographs (words that look the same but mean different things).
- It helps to provoke both a closer reading of the text and a better understanding of what the text really means. To a large extent the process of doing Ahab is more important than what is actually produced.
- It is a way of identifying candidate classes and frameworks from the collection of noun phrases, while keeping them in the context of related noun and verb phrases.

- It allows the diagrammatic manipulation of fragments, to make the dependencies between elements of the text clearer.

Ahab can be applied to a variety of texts at various stages throughout the development process. It has various complementary uses, as follows:

- It has been designed primarily to aid the understanding of a written statement of requirements, to aid the discovery of inconsistencies and ambiguities.
- It can be used as a half-way house between reading some text and doing some modelling by helping to identify areas of interest in the domain and their relationships with each other.
- It can be used for validation. The way Ahab is used here is broadly similar to that above, but with the aim of validating a model against some initial statement of requirements. Rather than, 'I don't understand this piece of text well enough to construct a model of it', the thought that initiates this use of Ahab is more like, 'Does this part of my model really reflect that part of the text?'

A blanket application of Ahab to a text is likely to be time-consuming and unproductive, so a first step is to exercise some editorial control and filter out parts of the document that are either well understood already, or are irrelevant. The reason for constructing Ahab fragments is that the piece of text is particularly obscure, or bears some, as yet unclear, relation to another piece of text. The decisions involved in constructing a fragment should provoke consideration of the meaning of the text.

20.4 Applying Ahab to NIMWeC

As a starting point for our NIMWeC case study work, a mock 'Invitation to Tender' (ITT) was prepared by someone very familiar with the real factory. Initially, this document was all the information we had, and we used Ahab to decipher it and to prepare the overview and glossary given in Chapter 4.

The document exhibited many of the faults common to 'real' ITTs.

Jargon

The weaving industry, in common with any specialised domain, has a language all its own. Part II of this book deals with terms such as warp, weft, pick, shuttle, jacquard and shed. All these terms need to be identified and understood. We prepared the weaving glossary in Chapter 4 in order to do this for our own purposes.

Synonyms and synecdoche

The document uses 'piece' and 'ribbon' interchangeably, and also 'warp thread' and 'warp end'.

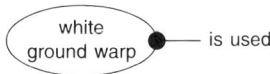

Figure 20.1 Unary fragment: *a white ground warp is used*

Sometimes 'loom' is used to refer to the whole machine, including control boxes, jacquards and the like; sometimes it used only for the part that does the weaving. Ahab helped us to discover this confusion, and we invented the term 'weaving frame' for the latter, to allow us to distinguish the part from the whole.

Incompleteness, and unnecessary detail

Despite a section entitled 'An introduction to narrow fabric weaving', the document assumed that we knew a lot more about this domain than was actually the case.

On the other hand, not everything it contained was relevant to the analysis. It went into asides about Swiss cuckoo clock manufacturers (who manufacture the patch panels), an Australian cub scout pack (who ordered a single 'name' to be that of the entire pack), and the selvedge pattern.

20.4.1 Constructing Ahab fragments

Ahab takes a simplified view of text analysis, taking parsing down to the level of the basic components of a clause. It considers only two basic components, roughly equivalent to noun phrases and verb phrases. Complex sentences have to be broken down into several simpler sentences of this form. For example,

> Each loom has a warp end breakage detector, which will stop the loom should a warp thread break (which happens every half hour or so).

becomes

> Each loom has a warp end breakage detector. A warp end breakage detector stops the loom should a warp thread break. A warp thread breaks every half hour or so.

Each phrase or clause is represented as a 'blob and line' fragment. Shaded blobs and dashed lines may be used to indicate similarities between one part of a diagram and another. For example, dashed ovals are used here to highlight things to do with time.

A *unary fragment* consists of a single noun phrase and a verb phrase (Figure 20.1).

A *binary fragment*, the usual case, consists of two noun phrases related by a verb phrase or conjunction (Figure 20.2).

A *ternary+ fragment* consists of any extension of a binary fragment, for example, a binary fragment related to a noun phrase, a clause or another fragment by some

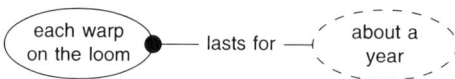

Figure 20.2 Binary fragment: *the warps on the looms last about a year at a time.* (Note that the phrase has been made singular.)

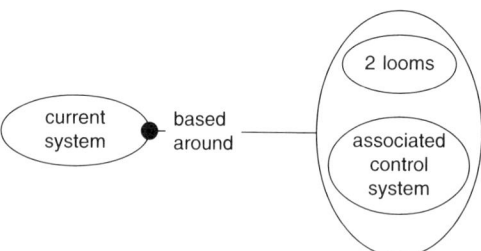

Figure 20.3 Ternary fragment: *The client's current system is based around two looms and the associated control system.*

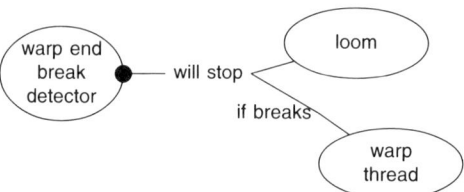

Figure 20.4 Ternary fragment: *A warp end breakage detector will stop the loom should a warp thread break.*

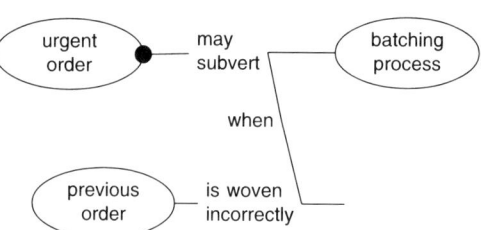

Figure 20.5 Ternary fragment: *An urgent order may subvert the batching process when a previous order has been woven incorrectly.*

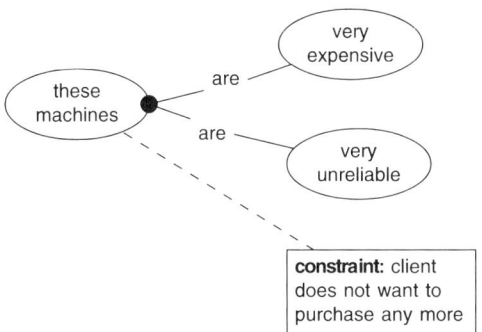

Figure 20.6 Annotated fragment: *These machines are very expensive and very unreliable. The client does not want to purchase any more of them.*

connector attached either to a noun phrase or the fragment as a whole. There are many possibilities (Figures 20.3, 20.4, 20.5).

20.4.2 Annotating fragments

Fragments may be annotated (Figure 20.6). This is usually done where a piece of text that relates to another piece of text seems to fall unproblematically into a general analysis category. Some categories that prove useful are *constraint*, *problem* and *change*. Other kinds of categories might be appropriate, depending on the context in which Ahab is being used.

20.4.3 Building an Ahab diagram from fragments

Individual fragments typically share some domain of discourse, so the next activity is to identify what refers to what by merging and overlapping fragments. This is intended to reveal the nature of the relationships between parts of the text. In an intractable text, the result may be questions about the domain to which the text refers.

With *overlapping*, two or more different noun phrases that refer to the same thing are overlapped (Figure 20.7).

With *merging*, two or more identical noun phrases or pronouns that refer to the same thing are replaced by a single noun phrase. One obvious candidate for merging is those complex sentences that were split into simpler sentences in order to make the fragments. With more experienced use of Ahab, the complex merged fragment can be drawn immediately. For example,

> Each loom has a warp end breakage detector, which will stop the loom should a warp thread break (which happens every half hour or so).

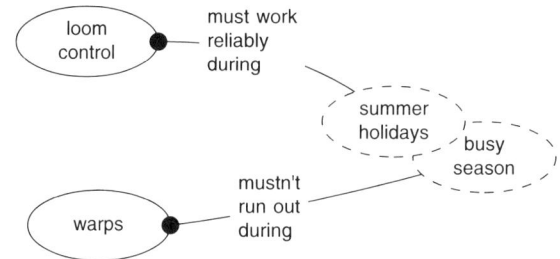

Figure 20.7 Overlapped fragments containing *busy season* and *summer holidays*

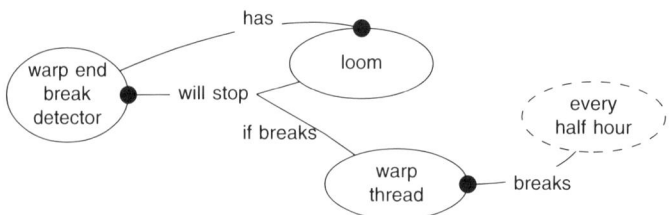

Figure 20.8 Merged fragments

is broken down to

> Each loom has a warp end breakage detector. A warp end breakage detector stops the loom should a warp thread break. A warp thread break every half hour or so.

and the whole is merged in Figure 20.8.

As more fragments are merged, a large Ahab diagram builds up. Related concepts should be placed close together if possible. For example, in Figure 20.9, things to do with timing constraints have been clustered.

Ahab is essentially an informal notation, and its purpose is to aid understanding. So if Ahab as presented here cannot clearly represent a particular piece of text, it should be modified or extended until it can.

20.5 Outputs from Ahab

For complex documents with a lot of new terminology and jargon, Ahab can be useful for determining a well-defined vocabulary. The merging and overlapping process can be used to develop a glossary of preferred terms, and to highlight unnecessary synonyms and inconsistent usage in the original text. Such a glossary is a valuable input to the rest of the analysis process.

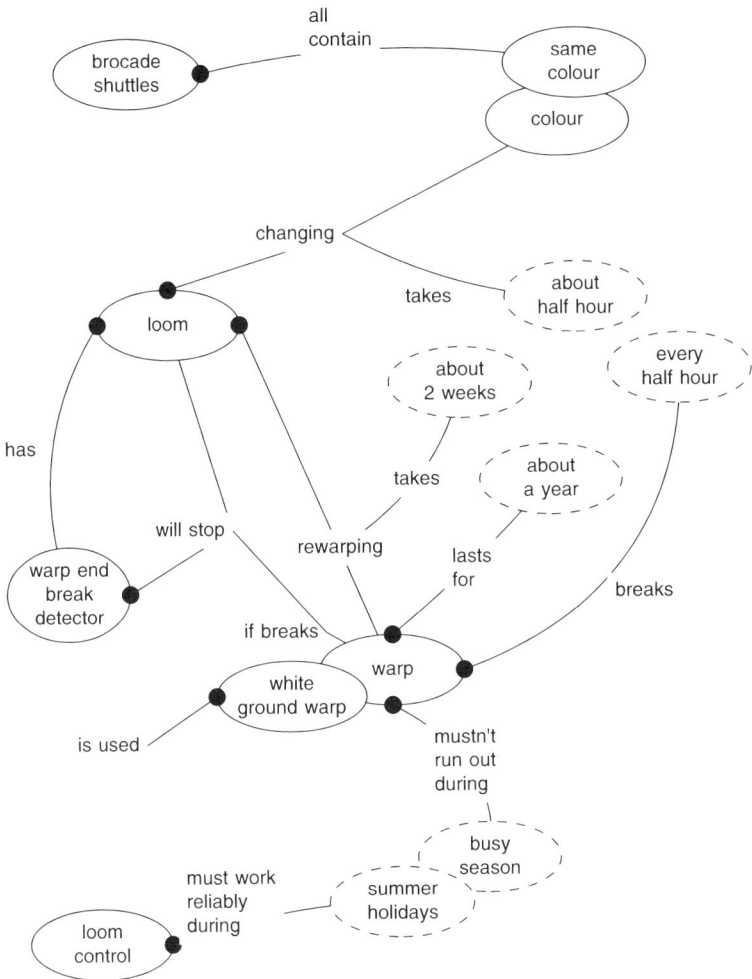

Figure 20.9 A large Ahab diagram of NIMWeC

The major output from Ahab is a better understanding of the text. The process of using Ahab—questioning the meaning of a piece of text—is more important than any Ahab diagrams that are produced. In fact, you might never get around to scribbling Ahab diagrams; after a while you might only need to 'think Ahab' when reading balderdash.

If necessary, Ahab can be used to produce a better version of a document. The consistent vocabulary and better understanding will help in writing a clearer document. The larger Ahab diagrams show clusters of related concepts, which may suggest a better structure for the new version.

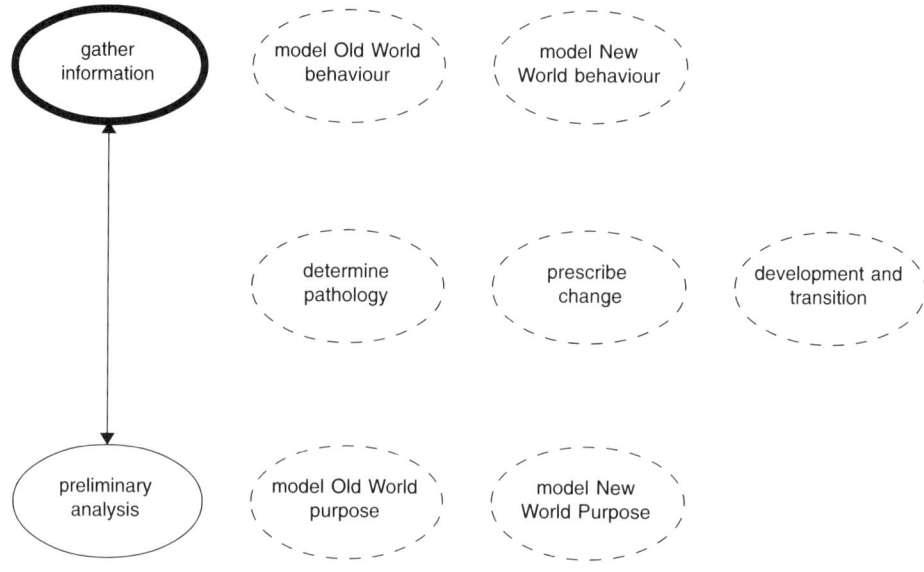

Figure 20.10 A process model for Ahab

20.6 Adding Ahab to the analysis process

The analysis activities supported by Ahab are highlighted in Figure 20.10. Ahab can be used at any stage of analysis, whenever the analyst is faced with obscure text. It is particularly relevant if there is limited access to the authors, to the client or to domain experts. Ahab is a technique for analysing natural-language text; it is *not* a modelling language like Grampus or Beluga.

The exact way in which Ahab is used depends on a number of factors, as follows:

- The form of the source material. Ahab has the most to offer for complex text.
- Familiarity with the problem domain. Where a lot is known about the problem domain, Ahab is of less utility.
- Presence of domain expert. The presence of a domain expert can either speed up the use of Ahab, with the potential for immediate clarification, or remove the need for Ahab altogether, with the domain expert taking over the role of information source.
- Availability of information. When responding to an Invitation To Tender, for example, source text might be all that is available, and Ahab could be of use. However, once a project is initiated, access to domain experts might be provided, with the original text assuming a secondary role.

These factors may determine when and where to use Ahab. Once a decision has been taken to use Ahab, the following points have to be considered:

- Editorial control and organisation of source material.
- Translation: how to turn source text into Ahab fragments (in particular dealing with complex sentence structures); how to cluster these; recognising commonly occurring patterns, such as ordering and conditionality; annotating problems, constraints, and so on, associated with an Ahab fragment.
- Traceability: Ahab is likely to 'sit between' source documents and ORCA models. Traceability is concerned with relating Ahab fragments back to their source and forward to those models that are based on them. Simple annotations allow, in principle, the connection to be made between source and Ahab. Possibilities exist for more sophisticated, perhaps tool-based, schemes for ensuring traceability.
- Version control: tracking updates in response to the acquisition of new or revised source material.
- Tool support: both version control and traceability would benefit from tool support.

As with the ORCA modelling languages, it is important that a technique such as Ahab is the servant of the analysis process, not its master. The ultimate goal for the analyst is always to improve understanding and draw valid conclusions.

Part V

Wider Issues

Chapter 21

Putting together an analysis project

21.1 Planning, monitoring and estimation

The activity of requirements definition has a tendency to be cut short in a rush to write some code. The analyst is sometimes the last person to hear about a development project, by which time it is often 'too late for all this analysis stuff'. Ideally, though, the requirements definition activity should have a generous *time allowance*, even at the expense of subsequent activities. Proceeding into IT system development with incomplete or ill-defined requirements is asking for trouble ('fools rush in where analysts fear to tread...').

The requirements definition activity should have explicit *deliverables*. It is important that the results of analysis are accessible to other parties (the client, the development team). The contractual status of a statement of requirements will normally need to be agreed with the client prior to IT system development.

The requirements definition activity should be *visible*, both within the analysis team and to the client. Frequent reviewing and validation are necessary to prevent analysis heading off in the wrong direction. Periodic milestones within the analysis activity are a good idea.

The requirements definition activity involves constructing models of the client's world and using these models to produce and support a statement of requirements. Requirements models also help an analyst (and a client) to understand the nature and complexity of the problem. This understanding can be used to influence the *planning and estimation* of subsequent development work. It follows that, ideally, a second round of planning and estimation should take place once requirements definition has been completed.

Personnel and organisational issues need to be addressed. Is there one analyst or a team of several? If there are several, how is the analysis activity partitioned and how do the analysts communicate? Is there any contact or continuity between the analysis team and the development team?

21.2 Validation

The importance of validating requirements and their supporting models cannot be stressed too much. Firstly, analysts may well be unfamiliar with a client's world, so the possibility of misunderstanding is considerable. Secondly, the consequences of misunderstanding at such an early stage of the process are potentially very serious.

There are two basic validation techniques:

- inspection and review
- demonstration and evaluation

With either technique, some form of *client participation* is essential. This can be obtained, for example, by client participation in reviews, or by obtaining feedback on the use of prototypes. However, it is important that the rules for client participation are clearly defined. It is generally more useful to get definite responses to specific questions, rather than to ask 'what do you think of it?' In particular, changes of fact (that is, corrections) should be distinguished from changes of requirement; the latter may need to be controlled in order to maintain some stability.

Prototyping involves implementing a mock-up or skeletal version of the delivered IT system. Its most important use is to validate known requirements and elicit further requirements.

A danger of using prototypes is that the client succumbs to 'love at first sight', and agrees to requirements that more detached consideration (or eventual use of the IT system) shows to be unsatisfactory. There is also a danger that the client asks for the prototype to be made faster, or more robust, or generally improved so that it can serve as the delivered system. This should be resisted, strongly! A true prototype should be 'thrown away' once it has served its purpose. Incremental development—a strategy that OO encourages—is often a good idea, but it should take place within the main development activity, not as part of the requirements definition.

Prototyping should be distinguished from *animation* of requirements models. Animation takes models that have a time dimension, as Beluga frameworks do, and unfolds the models in real time. Animation is most effective when the behaviour described by a model can be explored interactively, allowing a validator to ask questions such as 'what happens if I do this...?' or 'can I get into this (desirable or undesirable) state?' Animation requires tool support that is tailored to a particular notation, such as Beluga, and has a sophisticated user interface that presents behaviour and allows a validator to interact with the model. For the ORCA notations, animation is a topic that remains to be investigated.

21.3 HCI requirements

Requirements concerning Human Computer Interaction (HCI) are important for many applications. In some domains, such as Air Traffic Control, a major reason for IT system development is to alleviate ergonomic problems.

Pursuing the example of Air Traffic Control, it is clear that we can model the application domain: planes, flight plans, air corridors, and so on. However, these basic abstractions may already be defined for us by standard operating procedures. In this situation, defining requirements involves modelling the tasks performed by Air Traffic Controllers, and the processes by which they manipulate representations of planes, flight plans, and so on. A model of end-user tasks can then be used to support a statement of requirements, on the basis of which particular representations can be designed. Time and capacity requirements can be expressed as annotations to a task model.

The ORCA modelling languages can be used for modelling HCI domains, and expressing HCI requirements, as well as for describing application domains. In some cases the HCI requirements may be the major component.

Prototyping is often an effective way of determining HCI requirements. However, it is important that use of HCI prototypes seeks to elicit definite operational requirements rather than just demonstrating a particular 'look and feel'.

In defining HCI requirements, it is advisable to have access to potential users of the delivered system, or at least to someone who can represent their views. Information gathering from potential users could involve observing the performance of current or experimental tasks, as well as direct interviewing.

A problem with HCI requirements is that they are often expressed in a vague, if not vacuous, form: 'user friendly', 'suitable for untrained staff', 'consistent style', and so on. One of the tasks of an analyst is to make such requirements precise and define them in terms of a model of system use. Any requirement which cannot be defined in terms of a model must be treated with suspicion. For example, if a system needs to be 'suitable for untrained staff' it might turn out that there are essentially two modes of use, only one of which is relevant to 'untrained staff'. A response to this requirement would be to allow functionality associated with the 'expert' mode to be hidden from the general user.

21.4 'Non-functional' requirements

So-called *non-functional* requirements concern properties such as security, reliability and performance. ORCA handles this kind of requirement in two ways. Firstly, in purposive models the service descriptions (reliances and guarantees) can have *qualifiers*. These are typically adverbs—'handle payment *promptly*', 'deal with requests *fairly*', 'deliver items *reliably*'—although more complex phrases can be used.

These service description qualifiers allow informal expression of 'non-functional' requirements.

Subsequently, ORCA behavioural models should be used to make these qualifiers more precise. Timing requirements can be defined with reference to events within the dynamic behaviour. Fairness and reliability requirements can be expressed as statistical properties over the history. In some cases, specialised non-ORCA models (statistical or mathematical models) might need to be used to support the definition of quantitative requirements.

Capacity requirements

Beluga uses a cardinality construct to describe sets of objects, associations between objects, and event composition. This construct expresses an allowable range of values, which is usually one of the standard ranges (1, ?, *, +), but in general can be any finite or infinite set of integers. In addition, the typical or desired frequencies of subranges can be given, either formally (95%) or informally ('nearly always'). There is thus considerable scope for expressing quantitative requirements in an analysis model.

Timing requirements

Beluga provides a good basis for dealing with timing requirements, since the framework construct explicitly describes dynamic behaviour. It is relatively simple to annotate framework dynamics with timing constraints or required frequencies. Complex distributions of timing values could be referenced as external models. A more formal treatment of real-time properties is being investigated.

Reliability requirements

An ORCA purposive model looks at the provision of, and reliance on, services. This provides a focus for examining reliability requirements. For example, one might ask 'what is an adequate level of service?' or 'what aspects of server behaviour contribute towards guaranteeing adequate service?' It may be necessary to be more precise about the nature of co-operations at the purposive level, as well as describing required properties at the behavioural level. In some cases, we may want to model 'error' behaviours explicitly, where these are an unavoidable 'fact of life', such as failure of a communications link or hardware component. Beluga provides a special STOP event to represent failure of a behavioural strand. Having modelled error behaviours, we can then talk about the maximum acceptable frequencies with which they happen and specify what recovery behaviour is needed when they do happen.

Security requirements

A system might have a security requirement of the form 'an operation o on item i can be performed by user u, only if u has clearance to do o on i'. Such requirements may be captured using assertions within framework dynamics. An assertion can refer to the status of participants or to the prior occurrence (or non-occurrence) of other events. Where the conditions involve complex bodies of rules, a specialist model could be referenced. For example, a security policy may be expressed as a mathematical model in a formal specification language such as Z [Spivey 1992], [Barden et al. 1994].

Standards and statutes

We may have requirements that reference documentary definitions such as company procedures, legal statutes (for example, tax law) or technical standards. Such requirements are generally of the form: 'the system must conform to X', where conformance needs to be checked at some point in the development process. In some cases, it may be decided that such definitions are within the scope of the analysis and should be modelled. This may bring to light inconsistencies or incompatibilities with the New World, for example in company procedures. In most cases, however, it is probably sufficient for the statement of requirements merely to reference the definitions. In the future, standards might be expressed in such a form that they can be manipulated directly within an ORCA model.

Software properties

It is common to encounter requirements for properties of software (rather than complete systems), such as *extensibility*, *maintainability* and *portability*. These are really considerations for the development activity rather than requirements to be defined as a basis for development.

For example, the design and implementation authorities might wisely decide that the goals of extensibility and maintainability are best served by using an object oriented approach throughout the development, and (perhaps less wisely) that portability is maximised by using C++ on a Unix platform. They may be asked to justify their decisions, but the properties cannot actually be tested. This kind of requirement is not handled explicitly by ORCA.

Development constraints

In addition, there may be constraints on the development activity itself. For example, there may be requirements that the developed system interfaces with existing software or machines. This is really a constraint on the scope of change available to a system developer, and has implications for software design.

A client may make arbitrary or political choices of target technology ('it's got be PaneWare 6.2 on a HAL 2001'). These may have a significant impact on subsequent design choices.

Finally, there may be constraints on cost or timescale, which affect the nature and scope of any developed software.

21.5 Configuration management

Requirements and their supporting models are, in the nature of things, highly changeable. Firstly, an analyst's (and indeed, a client's) understanding of a domain develops as analysis proceeds. Secondly, the requirements being formulated become, one hopes, increasingly precise (and may even change radically). Thirdly, requirements definition often uses information gathered from multiple sources within a client organisation. Information from different sources may well be inconsistent, or may express conflicting requirements. Producing a single corpus of requirements and supporting models involves multiple cycles of revision and integration.

On the other hand, where the results of requirement definition are to form an agreed basis for system development, it is important that they are preserved, that their status is recorded, and that any subsequent change is controlled.

It follows that an effective Configuration Management (CM) regime is an important part of the overall process. A CM regime needs to address the following issues:

- what bodies of information constitute configuration items
- how configuration items are identified and stored
- how relationships between items are recorded, and how composites are defined
- the lifecycle of an item: the statuses and transitions between them
- change control procedures (request, access, authorisation)
- use of CM tools

Defining an appropriate CM regime needs to take into consideration:

- the scope and complexity of the analysis domain
- whether additional non-ORCA models are being used (see 21.4)
- whether there are single or multiple sources of information
- whether there are one or several analysts involved

It is clearly important not to burden the process with a CM regime that is too onerous. An analyst needs some freedom to work 'on the back of an envelope' (more probably, on the front of a whiteboard). On the other hand, it may be better to have a consistent CM regime right through a project, even if this seems excessive in the early stages.

21.6 Tool support

21.6.1 Roles for tool support

There is a prevalent view that tool support for methods is 'A Good Thing' and is therefore necessary for serious use of a method. However, while the scale and complexity of analysis products (in a particular situation) might require the use of tool support, there is nothing in ORCA that makes tool support essential. In fact, it is a matter of principle that the method should be usable, to some useful degree, with only paper or whiteboard.

To assess the need for tool support, we need to consider what tools might actually be used for.

Recording models in electronic form

There are obvious benefits in maintaining information in electronic form—it is then easily modifiable, reproducible, communicable, and so on. Both text and diagrams are amenable to electronic storage. Various standard tools can be used for this purpose, such as editors, drawing tools and filecard packages. These are becoming increasingly sophisticated and interlinkable.

Drawing diagrams

Given that we want to represent models in diagrammatic form, and hold diagrams electronically, there is an obvious requirement to support the drawing of diagrams. The diagrammatic forms of ORCA models are often complex, and the ability to adjust layout is important. In addition, there are standard shapes and symbols that may occur many times within a diagram, so the ability to duplicate and modify these elements is also desirable. Finally, it is often necessary to produce high quality diagrams for inclusion in delivered documents or live presentation. However, it is not clear that we need specialist CASE tools for this purpose since the latest drawing tools have increasingly powerful facilities (for example, with regard to layout), and considerable scope for customisation (for example, defining symbols, styles, and constraints).

Viewing complex models

ORCA models can become complex and multifaceted. There is thus a need for facilities to browse models, search for particular elements, provide alternative views, and provide filtered views in which only certain kinds information are shown. A problem with viewing complex models is the difficulty of providing an overview of a whole model and some notion of context within it. Without this capability, it is possible to become lost within a model. Although sophisticated navigation facilities (for example, hypertext style navigation) is desirable on usability grounds, it may create problems through lack of context and overview capability. Existing method support tools are still fairly primitive in this respect.

Checking models

Models constructed using any well-defined language are amenable to checking. Most obviously, models can be checked for correct syntax (see Part VI for definitions and explanations of the ORCA modelling languages). Syntax checking ensures that models are structurally well formed, although not necessarily meaningful. Checking static semantics ensures that the usage of names and types is correct (for example, if a name is used as a reference, then it must have been declared somewhere else). Although syntax and static semantics can be checked in 'batch mode' (as in a compiler for a programming language), it is more useful for an analyst to be able to check these interactively as a model is constructed since this is part of the process of making descriptions precise.

Within ORCA models there are various validity checks that can be performed. What these detect are not faults in the construction of models, but are consequences of the situation that the models describe. For example, in a Grampus model we are interested in detecting faulty co-operations, that is, guarantees or reliances for which only extrinsic descriptions exist. Such faulty co-operations are a common pathology (see Chapter 15 for examples). There is a similar, but more complex, set of validity checks applicable to Beluga models. It would be highly useful for detection of such conditions to be done by a tool rather than by inspection of a model by an analyst.

Animation

Animating models (where these have a time dimension) positively requires tools support if it is to be done; it is simply not feasible to conduct animation manually. Animation involves an analyst setting up scenarios from a model and then generating behavioural traces from these, making choices interactively where necessary. Alternative scenarios and choices can be explored. For this activity to be manageable, a tool needs a sophisticated user interface with regard to both presentation of behaviour and control of the animation. It is probable that an animation tool would need to be method specific since much depends on the underlying semantics of the modelling languages (in other words, what it is that a given model allows or disallows).

Configuration management

Configuration Management (CM) is discussed in Section 21.5. Handling versions, variants, change histories and access control is sufficiently tedious and exacting that it demands tool support. There are various CM tools available. However, these are intended for textual files (documents, program source code) or binaries rather than diagrams. One of the problems for the analyst is to determine the units of configuration management. Although we have tended to talk about 'models', what we typically have are interrelated fragments of models—in practice, we might never have a single monolithic 'model'. The problem is compounded if we do not construct our models from scratch. Given a library of reusable model fragments, we might wish to express derivation relationships with respect to already existing fragments.

One of the roles of CM support is to provide the analyst with an *audit trail* of the analysis process, recording raw information and models at different stages of analysis. In addition, the analyst might want to record alternative models, choices between these, and justification for the decisions made. In principle, an entire *rationale* for development could be constructed explicitly. This goes well beyond the capabilities of existing CM tools.

Supporting the process

It is somewhat ironic that 'method support tools' rarely support a *method*—prompting the performance of particular activities at appropriate times, providing guidance on what to do, helping resolve problem situations, and so on. It is much more common for tools to support modelling languages, and only implicitly the activity of model construction.

It is therefore worth distinguishing *process support* from language support. One reason why the former is a rare beast is that it is difficult to support processes (particularly analysis processes) in a sufficiently flexible manner. As earlier chapters have made plain, it is essential that any analysis method allows variants of the basic process to be used, as appropriate, in different situations. If a tool supports only a basic process, and does not allow an analyst to tailor the process, it is inevitably too constraining and will prove unsatisfactory.

Another difficulty with process support is that, to be effective, it requires a considerable degree of 'intelligence' from the tool. If the analyst asks: 'I'm in this situation now... (review of progress so far)... what's wrong? what do I do next?', what kind of useful response could a tool produce? Realistically, the analysis process is too flexible and too informal to be amenable to this degree of automation.

21.6.2 Advice to the unwary

Finally, the following pieces of advice are offered:
- Think of the reasons why you want tools (see the preceding section for possible roles for tool support). What are the benefits and costs?
- Consider whether you can just use general purpose tools. Can these be adapted or customised for your purposes? Can you configure 'meta-tools' (for example, tools that can be instantiated with language definitions)?
- Don't start using tools too early in the process; the whiteboard is probably the best tool for the early stages of analysis.
- A rough diagram that's right is better than a pretty one that's wrong.
- Using any kind of tool support is no substitute for thought.

Chapter 22

Life after ORCA—onward into development

22.1 Design and implementation considerations

Performing an ORCA analysis provides a context, a rationale, and a specification for proposed changes to a system. Some of these changes involve the development of IT components. These components are now the primary focus of attention.

In the NIMWeC example of Part II, the analysis led us to a prescription that involved the development and installation of a new control system for the looms. The new control system should:

- allocate batches to looms (in an 'intelligent' manner)
- translate batch details (item names, quantities) into loom instructions
- control the loom machinery via the electromechanical actuators and sensors
- provide an interactive interface to the operator, which is to:
  - allow entry of batch details
  - provide information on the status of batches and looms
  - allow manual intervention (removing a batch, taking a loom in/out of service, and so on)

The analysis models give us a precise understanding of these requirements. If necessary, a purely textual statement of requirements could be prepared using the outputs of analysis. We now need to move into the development activity.

The new control system will involve a mixture of software and hardware. Assuming that appropriate hardware components can be purchased, our task is to develop suitable software.

Computer software deals not only with the application domain, but also with various *implementation domains*. Domains can depend on the services provided by other 'lower level' domains. For the NIMWeC control system we might have the domain structure shown in Figure 22.1. The domains are shown as blobs with

22.1 Design and implementation considerations

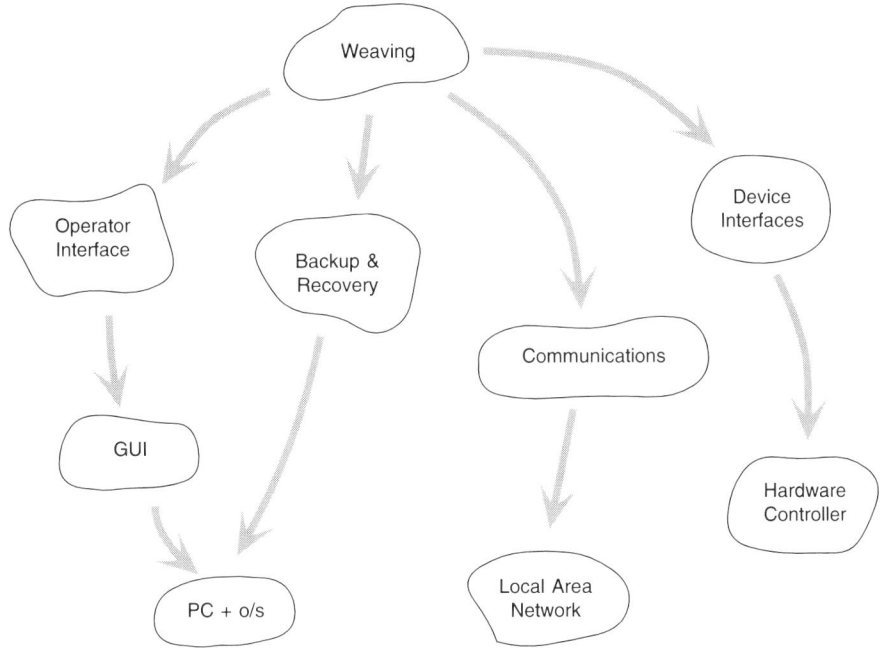

Figure 22.1 Implementation domain structure for NIMWeC

dependency arrows going from higher level to lower level domains (this informal notation is not part of an ORCA modelling language).

The Weaving domain has been investigated in our ORCA analysis—it deals with batches, items, looms, instructions, and so on. For the purposes of implementation, we need to go into more detail (for example, concerning the form of instructions), but the main concepts have been identified.

Invoking and controlling the functionality of the Weaving domain depends on the Operator Interface domain. This deals with operator sessions (log-on, log-off), modes of use, and the command 'language'. This domain depends in turn on the Graphical User Interface (GUI) domain, which provides user interface components (windows, dialogue boxes, menus, and so on). Typically, we would produce these by specialising and combining 'bought in' library classes. These classes depend ultimately on, say, a PC running a suitable operating system, but we want as little to do with this domain as possible.

In addition, there are domains that deal with communications, interfaces to hardware devices (for example, the electromagnets), and backup storage and recovery. These, too, depend ultimately on hardware domains.

We can now consider the issue of *reuse*. This is unlikely to be feasible in the top level Weaving domain, unless we have done similar developments for other weaving companies in the past. In contrast, for the lower level domains we should be able to use 'bought in' components with standard interfaces. A particular case is the use of a commercially available class library as the basis for implementation of the Graphical User Interface domain.

In the intermediate level domains, the feasibility of reuse is less clear. For example, in the Operator Interface domain, the concepts of User, Session, Command, and so on, are widely applicable. We might therefore have reusable *implementation frameworks* (sets of co-operating classes) that can be adapted for our particular development. Such reusable implementation frameworks would have come out of a variety of previous developments, and would be maintained within an 'in house' software library.

The *scope for OO development* varies between different domains. We would certainly like to use OO in the implementation of the Weaving domain in order to make maximum use of our analysis models. However, we should be careful to consider the classes involved from an implementation perspective rather than from an analysis perspective. For example, in our analysis model each *OrderItem* in a *Batch* is associated with a set of *Instructions*. In our implementation, on the other hand, the instructions for an entire batch might be sent from the operator's terminal to a hardware controller (in a numerical encoding) when the batch is about to be woven. We could thus have a *batchInstructions* feature on the Batch class, containing a numerical representation of the instructions for all items in a batch, and not have an Instruction class at all. It is important to realise that there is no 'handle-cranking' way of turning analysis models into good designs.

The Operator Interface domain is another candidate for OO development, since it sits between the Weaving domain and the Graphical User Interface domain. The latter is almost necessarily object oriented (there are now many GUI class libraries).

In contrast, the lowest level domains may not be amenable to OO development. Accessing communications facilities may involve calls to procedural code; programming of device controllers may need to use low level languages. Thus we cannot expect to take a universally object oriented approach to development.

22.2 Design methods

This leads us on to consideration of *OO design methods*. Although some of these methods are billed as 'analysis and design' methods, the analysis activity is generally rather limited: we advocate using ORCA instead. As regards the design activity, different methods provide the following different kinds of support:

- Notations for design. [Booch 1991] is a good source of notations and textual templates.
- Heuristics for design. The Class–Responsibility–Collaboration approach of [Wirfs-Brock et al. 1990] is well-known and helpful.

- Consideration of architecture and design issues (such as concurrency, data storage, resource requirements). The OMT method of [Rumbaugh et al. 1991] provides advice in this area.
- Development processes (see comments below).

The methods of [Martin & Odell 1992], [Jacobson 1992] and [Goldstein & Alger 1992] should also be mentioned.

ORCA's Beluga language for behavioural modelling has much in common with existing OO design notations, particularly as regards static structure (classes, features, associations, generalisation/specialisation). However, the approach to dynamic behaviour is rather different. For analysis, we want a language that allows us to express patterns of behaviour, and compose patterns into larger patterns, without worrying about the details of interaction and ordering. For design, on the other hand, we are interested in building up a system by combining deterministic (usually sequential) components, and defining precisely how they interact and interleave (that is, the 'flow of control').

Beluga could be used for design purposes, provided that two points are recognised. Firstly, it is important to remember that analysis models are talking about the 'real world', whereas designs are talking about pieces of software. Secondly, in modelling for analysis the aim is to express only that which is essential; in design, the aim is to decide on definite mechanisms for information processing.

ORCA's Grampus language for purposive modelling has no obvious counterpart in existing design methods. An interesting possibility is that it could be used to express the ideas of the CRC method—responsibility, contract, collaboration [Wirfs-Brock et al. 1990].

In describing ORCA, we have emphasised the need for *process design*: an analysis process needs to be designed for the particular situation, integrating the technical approach with strategies for the wider project (see Chapter 21). A similar process design activity needs to take place for development. Suitable techniques, languages and notations need to be selected, perhaps from different methods. Strategies for validation and verification, documentation and configuration management need to be determined; planning and progress monitoring need to be addressed. Finally, the overall nature of the development process needs to be determined—whether the development is linear, incremental or evolutionary, the role of prototyping in the process, and the scope for software reuse.

Finally, we should discuss the relationship of ORCA to *non-OO methods*. ORCA covers much the same ground as traditional business analysis, but provides a more rigorous process and more powerful modelling languages. In particular, entity-relationship data modelling, as used for 'strategic data models', is subsumed by the Beluga language. Ideas from traditional business analysis, such as critical success factors, can be used within ORCA and tied in to ORCA models. Soft Systems Methodology [Checkland 1981], [Checkland & Scholes 1991] analyses organisations in terms of objectives, conflicts, mission statements and outlooks; this can be used as part of ORCA Preliminary Analysis. Non-OO methods such as [DeMarco

1978] could be used for design, following an ORCA analysis, but the continuity of approach would be lost. While, as mentioned earlier, there may be implementation domains for which OO development is not the most appropriate, OO is increasingly recognised as the principal approach for mainstream IT development.

22.3 The ORCA method

This book presents a set of ideas about analysing purposive systems, together with illustrations of how these ideas can be used. We do not regard the ORCA method as in any way the 'last word' on the subject. Quite the opposite: there is some way yet to go before we have adequate means to describe and understand complex systems. 'Methods' presented in books such as this are necessarily a compromise between pushing forward with new ideas and techniques, and holding back in order to maximise coherence and rigour. Thus, the ORCA method is by no means 'graven in stone', and we hope to see it evolve in the light of experience. It is not a religious dogma to which the faithful need adhere without deviation. Use of the method should be judicious and pragmatic. The aim should always be to assist understanding and help valid conclusions to be drawn: a method should provide a tool for thought.

Part VI

Appendices

Appendix A
Defining modelling languages

A.1 Introduction

In the case studies discussed in this book, parts of the Grampus and Beluga modelling languages are introduced by example, as and when they are needed. Complete descriptions of the languages are given in the following two appendices. These descriptions are provided not for the day-to-day user of ORCA, but for those interested in delving deep, for example in order to build tools. This appendix describes the notation used for those descriptions.

A.2 Syntax

Grampus and Beluga are defined in terms of their *abstract syntax* and various *concrete syntaxes*. The abstract syntax defines the underlying constructs of the language, while the concrete syntaxes provide ways of representing these constructs in a textual or diagrammatic form. [Meyer 1990, chapter 3] explains these ideas in more detail, and introduces a syntax-definition language called *Metanot*, which we use in a slightly modified form in the following appendices. In the descriptions given below, we also use Beluga diagrams to summarise the relationships between the abstract constructs.

A.2.1 Abstract syntax

Description

Each abstract syntax definition consists of a list of *productions*. A production has two parts: on the left is the name of the construct being defined, and on the right is its definition. A construct can be defined as a *choice* between other constructs, or as an *aggregate* of other constructs. (These meta-level terms should not be confused with the Beluga constructs called 'choice' and 'aggregation'.) For example:

$Choice \longrightarrow Choice1$
$\qquad\quad |\ \ Choice2$
$\qquad\quad |\ \ ChoiceN$

$Aggregate \longrightarrow terminal : \textsc{Construct1}$
$\qquad\qquad\ \ ;\ \ optional : Construct2^?$
$\qquad\qquad\ \ ;\ \ zeroOrMore : Construct3^*$
$\qquad\qquad\ \ ;\ \ oneOrMore : Construct4^+$

Each component of an aggregate construct has the following parts:

- A *name*, by which it may be referred to elsewhere, and which can be used to indicate its meaning informally. The names in the Aggregate example above are *terminal*, *optional*, *zeroOrMore*, and *oneOrMore*.
- A *type*, which is a construct name. A construct name written in SMALLCAPITALS indicates a terminal construct, not further defined.
- A *cardinality*. Various symbols may follow the type.
   No symbol (the default) indicates exactly one item in the component.
   ? A question mark indicates that the component is optional.
   * A star indicates a list of zero or more components.
   + A plus indicates a list of one or more components.

Components with the same type may be folded into a single line, if desired.

Example

In a programming language, a *Statement* might be defined using a choice construct as an *Assignment*, or an *IfStatement*, or a *ProcedureCall*, or a .... The *IfStatement* might be defined using an aggregate construct as having a test of type *Expression*, and a *thenBranch* and an optional *elseBranch*, both of type *Statement*. The *ProcedureCall* might be defined using an aggregate construct as having a name of type NAME (where NAME is a terminal construct, not further defined) and a *parameterList* consisting of zero or more components of type *Expression*:

$Statement \longrightarrow Assignment$
$\qquad\qquad\ \ |\ \ IfStatement$
$\qquad\qquad\ \ |\ \ ProcedureCall$
$\qquad\qquad\ \ |\ \ \ldots$

$IfStatement \longrightarrow test : Expression$
$\qquad\qquad\quad\ ;\ \ then : Statement$
$\qquad\qquad\quad\ ;\ \ else : Statement^?$

$$ProcedureCall \longrightarrow name : \textsc{Name}$$
$$;\ parameterList : Expression^*$$

A.2.2 Concrete syntaxes

Different concrete syntaxes

Grampus and Beluga have a variety of concrete syntaxes, which play different roles.

- The 'pretty' textual form contains special non-ascii characters (for example, various kinds of arrow). Ability to use this form depends on the available fonts. It can be used for presentation quality type-set documents and for freehand use.
- The 'basic' textual form is pure ascii (no symbols). It is intended for use on character-based terminals where only ascii characters are available, in particular for electronic mail and document interchange.
- The diagrammatic form is available for
  - exploratory use by the analyst
  - presentation of models to clients

Using diagrams

The diagrammatic concrete syntax described in the following appendices can be used piecemeal (for example, during exploratory analysis) or linked into large diagrams (for example, for discussing with other people). It does not have to be structured into 'X Diagrams', 'Y Diagrams', and so on. It is up to an analyst to determine structuring and layout conventions.

The diagrammatic syntax can (within reason) be extended with abbreviations, annotations or decorations to aid readability or to indicate the status of information (for example, the status of pieces of model). The corresponding textual form can be used to determine exactly what the diagrams mean.

Diagrams are not an end in themselves; don't get carried away with their production. Beware of producing 'pan-galactic' diagrams for a whole model, without specific benefits in mind. However, don't go to the other extreme and draw a diagram that is worth only a dozen words. A diagram should have a *macro-reading* (structural properties that are apparent when viewed from a distance) and a *micro-reading* (detailed information—textual, symbolic or graphical—that is apparent when viewed close up). [Tufte 1983] and [Tufte 1990] contain excellent discussions and examples of good diagrams, and some revealing examples of poor diagrams, too.

Concrete syntax definition conventions

Where there is a difference, both 'pretty' and 'basic' textual forms are given, in that order.

In the concrete syntax descriptions, embedded syntactic components are named in ⟨*bracketed italic*⟩, terminals are shown as literal text, or as the relevant diagram fragments. Multiple components are shown as

⟨*componentList*⟩ separator ...

if the list occurs on one line, or as

⟨*componentList*⟩ separator
...

if the separate items occur on separate lines. If a list of components is empty, brackets may be omitted.

Example

The concrete forms of the *IfStatement* and *ProcedureCall* described above might be

IfStatement ::= if ⟨*test*⟩ then ⟨*then*⟩ else ⟨*else*⟩ fi
 | if ⟨*test*⟩ then ⟨*then*⟩ fi

ProcedureCall ::= ⟨*name*⟩(⟨*parameterList*⟩, ...)

Actual instances of such statements (depending on the concrete definitions of *Expression*) might look like:

if a=b then skip else a:=3 fi
draw(x, y, str)
resetClock

A.3 Semantics

A.3.1 Static semantics

In order to keep the syntax as simple as possible, many consistency constraints are expressed as 'static semantics', so called by analogy to the static, or compile time, checks made on programming language texts. These consistency checks are expressed informally in the language description appendices. For example, 'class names are unique within a model'.

A.3.2 Dynamic semantics

The dynamic semantics, what a piece of model 'means', should be defined formally so that models can be manipulated, proved equivalent to one another, shown not to include deadlocking behaviour, and so on. Such a formal definition is beyond the scope of this book. The earlier case study chapters introduce the meaning informally, by example, and the language description appendices give an informal description of each construct's meaning with its abstract syntax definition.

Appendix B

Grampus—the Purposive modelling language

B.1 Introduction

Grampus (**G**uarantee **R**ely **A**pproach to **M**odelling **Pu**rpose in **S**ystems) is a language for modelling a system in terms of the arrangements for service provision. These arrangements are what gives a system its overall purpose.

Structurally, a Grampus model describes a *cluster* of co-operating *roles* (or, exceptionally, a single role in an undefined 'environment'). See Figure B.1.

A role is described in terms of the *services* it *guarantees* to provide to other roles within a cluster, and those it *relies* on getting from other roles in the cluster.

A *co-operation* is the matching of a service reliance of one role with one or more service guarantees of other roles. Describing the guarantees and reliances of one role implies guarantees and reliances of other roles. These implied service descriptions are said to be *extrinsic* descriptions of those other roles; the service descriptions that generate them are said to be *intrinsic* to the role.

A role may be analysed in terms of a *formation* into a *cluster* of subroles, and may *delegate* an outer service to its subroles' services. A role may be modelled by a number of formations.

Co-operations and delegations may be characterised as OKAY, PROBLEMATIC (mismatched) or UNANALYSED.

B.2 Roles and services

A Grampus model is either a cluster or a single role:

$$\begin{aligned} \textit{GrampusModel} &\longrightarrow \textit{Cluster} \\ &\mid \textit{Role} \end{aligned}$$

B.2 Roles and services

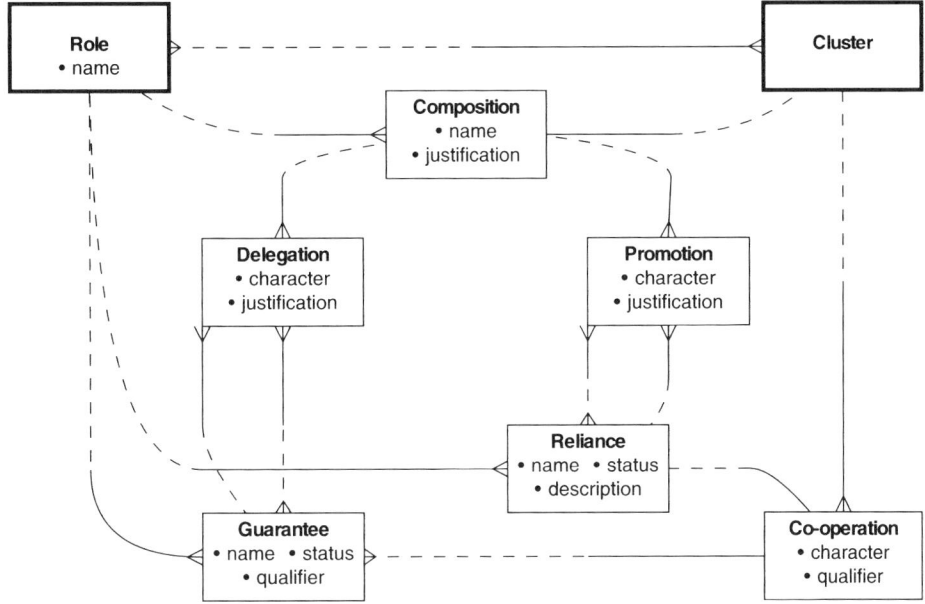

Figure B.1 The structure of the Grampus language

$Role \longrightarrow name : \textsc{RoleName}$
  $;\ guaranteeList : Guarantee^*$
  $;\ relianceList : Reliance^*$
  $;\ formationList : Formation^*$

A role has a *name*, and a description of various services *guaranteed* and *relied* upon. A role may have one or more *formations* that model its internal structure in different ways.

$Guarantee \longrightarrow name : \textsc{GuaranteeName}$
  $;\ qualifier : \textsc{Text}$
  $;\ status : Status$

$Reliance \longrightarrow name : \textsc{RelianceName}$
  $;\ qualifier : \textsc{Text}$
  $;\ status : Status$

A guarantee or reliance is characterised by a *name*, a textual *qualifier* and a *status* (intrinsic or extrinsic).

$Status \longrightarrow \textsc{Intrinsic}$
  $\mid\ \textsc{Extrinsic}$

244   Appendix B   Grampus—the Purposive modelling language

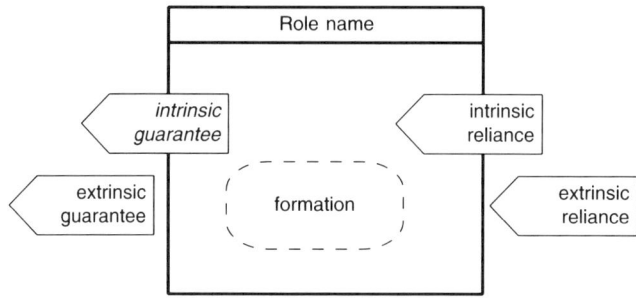

- RoleName: *intrinsic guarantee* and its qualifier

Figure B.2 Concrete syntax for Role

An intrinsic guarantee or reliance is part of a role's understanding of itself. An extrinsic guarantee or reliance is a consequence of an intrinsic guarantee or reliance of another role.

Static semantics

Guarantee and reliance names must be unique within a model, but note that the same role can appear in alternative formations. These names are 'short descriptions', not identifiers.

Concrete syntax

A role is drawn in a box with its name at the top. The guarantees and reliances are drawn as named 'pointed boxes'. Where a qualifier is brief, it may also be written in the box (after the name, in parentheses); otherwise it may be written as a footnote (in which case the name is italicised to indicate the existence of a qualifier). See Figure B.2. Boxes pointing out of the role represent guarantees (made to other roles); boxes pointing into the role represent reliances (required from other roles).

Guarantees and reliances drawn outside (next to) the role represent extrinsic descriptions; guarantees and reliances drawn intersecting the role box represent intrinsic descriptions.

Services may be drawn on left, right or bottom edges of a role box. A role may optionally be drawn with one of its formations shown (see later).

B.3   Cluster and co-operation

$Cluster \longrightarrow componentList : Role^+$
$; \ cooperationList : Cooperation^*$

A cluster is a collection of two or more *component* roles, and of *co-operations* between these roles, showing how relies of one are met by guarantees of others.

$Cooperation \longrightarrow reliance : (\textsc{RoleName}\ \textsc{RelianceName})$
$\quad ;\ guaranteeList : (\textsc{RoleName}\ \textsc{GuaranteeName})^+$
$\quad ;\ characterisation : Characterisation$
$\quad ;\ justification : \textsc{Text}$

A co-operation (between roles in a cluster) identifies a *reliance* of one of the roles, and the *guarantees* of other roles that satisfy this reliance. The co-operation can be *characterised* as OKAY, PROBLEMATIC or UNANALYSED. The justification may be used to explain why the guarantees are believed to satisfy the rely, or otherwise.

$Characterisation \longrightarrow \textsc{okay}$
$\quad\quad\quad\quad\quad\quad\quad\ |\ \textsc{problematic}$
$\quad\quad\quad\quad\quad\quad\quad\ |\ \textsc{unanalysed}$

Okay co-operations occur when an intrinsic reliance is believed to be satisfied by intrinsic guarantees. Problematic co-operations occur when this is not the case. Intrinsic–extrinsic co-operations are always problematic. The characterisation may require a justification.

Static semantics

A cluster must contain two or more roles.

In a co-operation, the reliance must be in the appropriate *role.relianceList*; each guarantee must be in the appropriate *role.guaranteeList*.

No co-operation is extrinsic–extrinsic. Every extrinsic reliance or guarantee must appear in some co-operation.

Concrete syntax

A co-operation is drawn in by connecting the relevant services with a line (straight or curved); the characterisation is indicated by the style of line (see Figure B.3). Note that UNANALYSED lines can be changed to PROBLEMATIC or OKAY lines, and PROBLEMATIC to OKAY, by over-writing.

Typically, okay co-operations are intrinsic–intrinsic, while problematic and unanalysed co-operations are intrinsic–extrinsic.

Where a co-operation involves multiple guarantees, a circular connector is used to join them. See Figure B.4.

Where a justification is brief, it may be written on a co-operation diagram (in a dashed or fuzzy round-cornered box, to avoid confusion with the role boxes, attached to a circular connector), otherwise it may be written as a footnote. See Figure B.5.

A cluster diagram is the composition of all its contained role and co-operation diagrams.

246   Appendix B Grampus—the Purposive modelling language

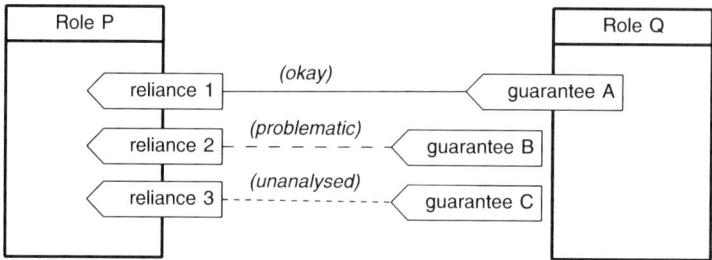

Figure B.3 Concrete syntax for Co-operation

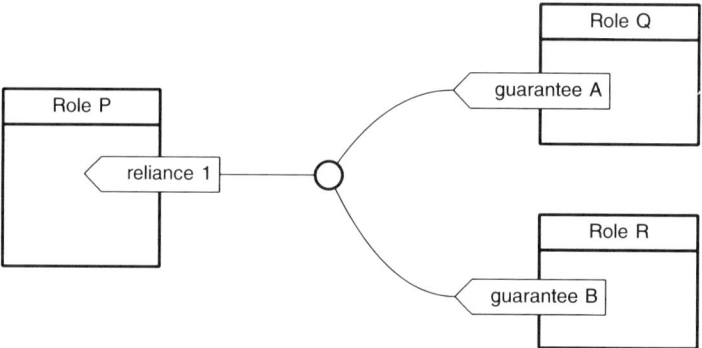

Figure B.4 Concrete syntax for multiple guarantees

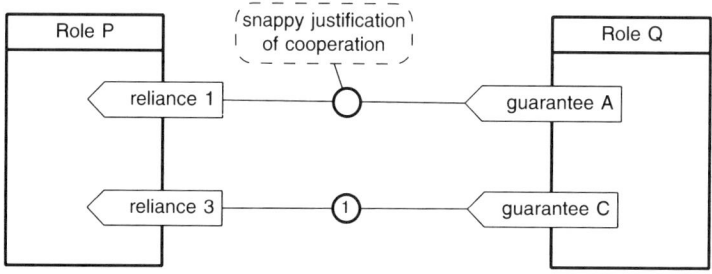

- (1) rather longer and more detailed justification of other cooperation

Figure B.5 Concrete syntax for Justifications

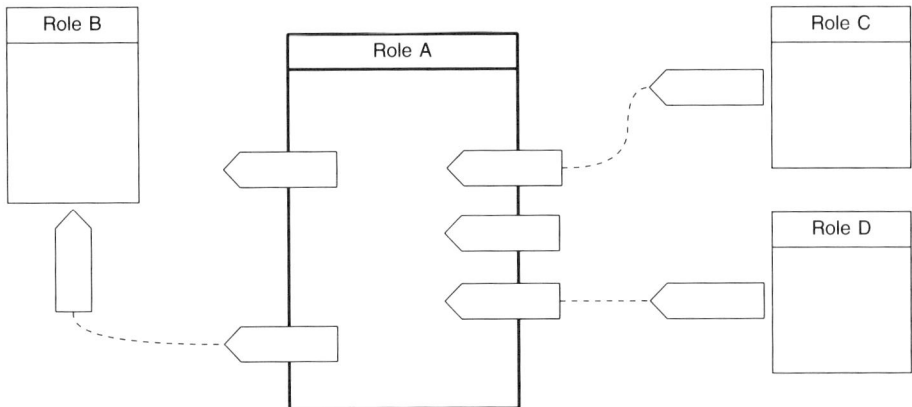

Figure B.6 The viewpoint of Role A

Where there is an information source for each role in a cluster, it may be useful (initially) to have a separate diagram for each 'viewpoint', where a viewpoint consists of a role and all its intrinsic reliances and guarantees, together with any intrinsic–extrinsic co-operations derived from these. For example, the viewpoint for some role A might look as in Figure B.6. Note that one of A's intrinsic guarantees is not involved in a co-operation; A has no assumption about which other role relies on it. Also, one of A's intrinsic reliances is not involved in a co-operation; A has no assumption about which other role guarantees it. Note also that the co-operations are shown as 'unanalysed'. When, in due course, multiple viewpoints are combined, the analyst will attempt to match intrinsics and extrinsics, linking roles with okay (intrinsic–intrinsic) co-operations, where matching is successful.

B.4 Formation, delegation and promotion

$$
\begin{aligned}
Formation \longrightarrow \ & name : \text{FORMATIONNAME}^? \\
; \ & cluster : Cluster \\
; \ & delegationList : Delegation^* \\
; \ & promotionList : Promotion^* \\
; \ & justification : \text{TEXT}
\end{aligned}
$$

A formation (of a role) can be *named*; where there are alternative formations, the names can indicate the grounds for the different analyses (for example, 'functional', 'organisational'). A formation consists of a *cluster* (the internal structure), a set of *delegations*, which delegate an outer guarantee to inner components, and a set of *promotions*, which promote inner reliances to an outer reliance. That this formation models the outer role may need to be *justified*.

$$
\begin{aligned}
Delegation \longrightarrow\ & outer : \text{GuaranteeName} \\
;\ & innerList : (\text{RoleName GuaranteeName})^+ \\
;\ & characterisation : Characterisation \\
;\ & justification : \text{Text}
\end{aligned}
$$

A delegation (of a guarantee of the outer role to guarantees of components in a formation) identifies the guarantee of the outer role and the guarantees of the *inner* component roles to which it is delegated. The status of the delegation can be *characterised*. That a delegation does satisfy the outer guarantee may need to be *justified*.

$$
\begin{aligned}
Promotion \longrightarrow\ & outer : \text{RelianceName} \\
;\ & innerList : (\text{RoleName RelianceName})^+ \\
;\ & characterisation : Characterisation \\
;\ & justification : \text{Text}
\end{aligned}
$$

A promotion (of reliances of components of a formation to a reliance of the outer role) identifies the reliance of the *outer* role and the reliances of *inner* component roles from which it is derived. The status of the promotion can be *characterised*. That a promotion does satisfy the inner reliances may need to be *justified*.

Static semantics

Alternative formations of the same role must have different names. Outer guarantees must be intrinsic. Each outer guarantee must be provided by an inner role.

Concrete syntax

One of a role's formations can be drawn inside the role box. The cluster is drawn as shown earlier. Delegations and promotions are drawn in the same manner as co-operations. If a role has multiple formations, the formation name is appended to the outer role's name, as *RoleName.FormationName* (Figure B.7).

The outer role box may be elided (Figure B.8).

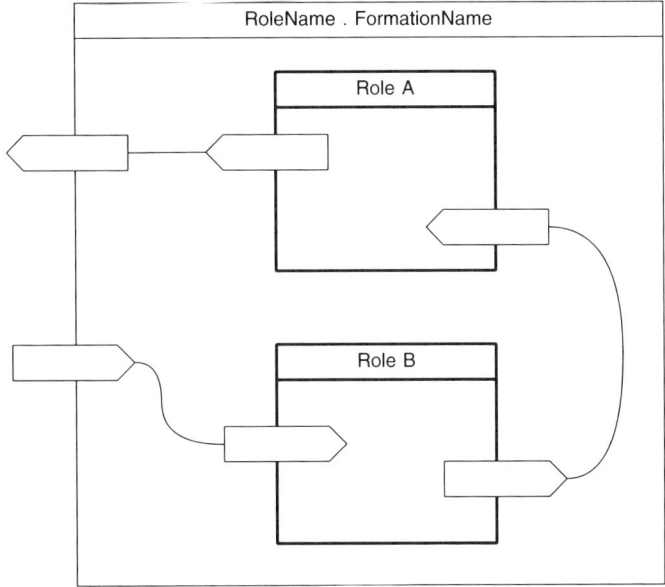

Figure B.7 Concrete syntax for Formation

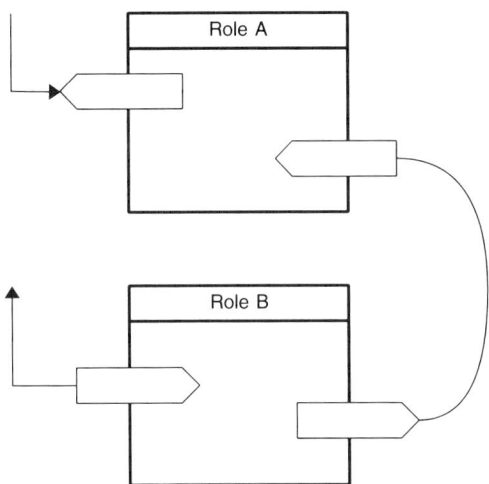

Figure B.8 Eliding the outer role box

Appendix C

Beluga—the Behavioural modelling language

C.1 Beluga models

Beluga (**B**ehavioural **L**anguage **U**nderpinning **G**eneric **A**nalysis) is a language for modelling a system in terms of the classes of object in the system, and the possible histories of objects and their interactions. The concepts and notations are introduced in Chapter 12, which should be read before referring to this chapter.

A Beluga model is a collection of class definitions and framework definitions.

$$BelugaModel \longrightarrow classList : Class^* \\ ; \; frameworkList : Framework^*$$

In terms of abstraction, the 'top level' of a model may be either a class or a framework. However, class definitions and framework definitions are syntactically disjoint—they are linked by reference (frameworks reference classes). The major 'name spaces' (for class/framework, constituent set, feature, association, local declaration and status) are distinguished; these have different types (CLASSNAME, FEATURENAME, etc.).

Beluga's textual concrete syntax (not used in the main body of the book) is defined along with its abstract syntax. Its diagrammatic concrete syntax description is gathered together in a single section at the end of this appendix.

C.2 Classes

The class is the object oriented building block. It has features that describe its abstract behaviour, and may inherit from (multiple) parents. Features involved in particular behaviours may be grouped into facets. See Figure C.1

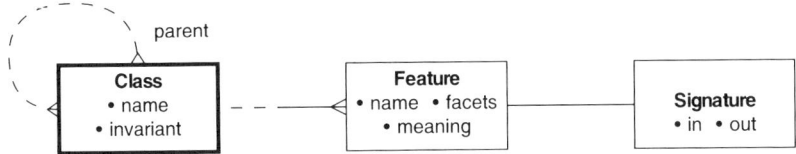

Figure C.1 Class, feature, and signature

C.2.1 Class

$Class \longrightarrow name : \text{CLASSNAME}$
    ; $parentList : \text{CLASSNAME}^*$
    ; $featureList : Feature^*$
    ; $invariant : \text{PREDICATE}^?$

A class has a *name*, and may optionally have a set of *parent* classes from which it inherits, a set of features (operations and attributes), and a class *invariant*.

Static semantics

Class names are unique in a model.

Concrete syntax

The keyword 'class' and the class name are written above a line. Below the line, the various components are listed in sections, with each section labelled by an appropriate keyword. The order in which these sections are written is not significant.

$Class ::=$

class	$\langle name \rangle$
parents :	$\langle parentList \rangle; \ldots$
features :	$\langle featureList \rangle;$
	$\ldots$
invariant :	$\langle invariant \rangle$

C.2.2 Feature

$Feature \longrightarrow name : \text{FEATURENAME}$
    ; $facetList : \text{FACETNAME}^*$
    ; $sig : Signature^?$
    ; $meaning : \text{MEANING}^?$

A feature has a *name*, an optional list of *facets*, an optional *signature*, and an optional *meaning*. The meaning may be informal text, executable code, pre and post conditions (using the tags rely: and guarantee:, and referring to 'after' values of attributes with a prime, v'), or empty (in which case the informal 'meaning' of the feature is indicated somehow by its name). The meaning can indicate that a feature is a *specialised* form of its parent feature, or *derived* from a feature of a related class.

Static semantics

Feature names must be unique within a class and all its parents, but the same name can be used for different features in different classes. Different classes may use the same facet name in order to indicate that a piece of behaviour involving (instances of) these classes occurs by invoking features with the commonly named facet.

Concrete syntax

$$Feature ::= \\ \{\langle facetList \rangle, \ldots\} \\ \langle name \rangle \langle sig \rangle \\ // \langle meaning \rangle$$

The facet braces are omitted if there are no facet names. The meaning comment slashes are omitted if no meaning is given.

C.2.3 Signature

$$Signature \longrightarrow ins, outs : FormalParams^*$$

A signature has two lists of formal parameters, indicating the formal *inputs* and *outputs*.

Concrete syntax

There is a variety of concrete forms, depending on whether the lists have many, one or no components. For example

$$\begin{aligned} Signature ::=\ & (\langle ins \rangle; \ldots) : (\langle outs \rangle; \ldots) & //\text{many in and out} \\ |\ & (\langle ins \rangle; \ldots) & //\text{many in, no out} \\ |\ & : (\langle outs \rangle; \ldots) & //\text{no in, many out} \\ |\ & : \langle outs \rangle & //\text{no in, 1 out} \end{aligned}$$

C.2.4 Formal parameter

$$\begin{aligned} FormalParam \longrightarrow\ & SimpleFParam \\ |\ & ClassFParam \end{aligned}$$

A formal parameter is either a *simple* formal parameter or *class* formal parameter.

$SimpleFParam \longrightarrow name :$ NAME
$\qquad \qquad \qquad \;\; ; \;\; type :$ SIMPLETYPE

A simple formal parameter has a *name* (so that it can be referred to in pre and post conditions, for example), and a *type* (which will be something simple such as 'number', not defined using a class).

$ClassFParam \longrightarrow name :$ NAME
$\qquad \qquad \qquad \; ; \;\; type :$ CLASSNAME
$\qquad \qquad \qquad \; ; \;\; card :$ Cardinality

A class formal parameter has a *name*, so that, for example, it can be referred to in pre and post conditions, a *type*, the name of the class that defines it, and a *cardinality*.

$Cardinality \longrightarrow range :$ Range
$\qquad \qquad \qquad \; ; \;\; frequencyList :$ Frequency*

A cardinality is a *range*, which says how many instances may be present in a parameter, and a set of *frequencies*, which indicate where in a range the number of instances is likely to lie.

$Range \longrightarrow nos :$ NAT$^+$

The range is a set of numbers. The actual number of instances on any occasion will be a member of this set.

$Frequency \longrightarrow proportion :$ TEXT
$\qquad \qquad \qquad \; ; \;\; subrange :$ Range

A frequency consists of a *proportion* of the time a certain *subrange* of number of instances actually occurs. This can be useful to indicate typical values when range is just 'any number'. There is no requirement for a distribution to be 'complete'. For example, we could say that the range is any number, where half the time there will be 0–3 instances, and 0.1% of the time there will be 1000 instances.

Static semantics

The sum of the proportions in a cardinality may not exceed one, and the subranges should be disjoint.

Concrete syntax

Simple formal parameters are written as name, colon, simple type, and class formal parameters as name, colon, class type and cardinality (either bracketted, or written as a superscript). A shorthand version is available for multiple parameters of the same type (and cardinality, where appropriate): the names may be folded into a single statement.

$$SimpleFParam ::= \langle name \rangle : \langle type \rangle$$
$$| \ \langle name1 \rangle, \ldots, \langle nameN \rangle : \langle type \rangle$$

$$ClassFParam ::= \langle name \rangle : \langle type \rangle (\langle card \rangle)$$
$$| \ \langle name \rangle : \langle type \rangle^{\langle card \rangle}$$
$$| \ \langle name1 \rangle, \ldots, \langle nameN \rangle : \langle type \rangle (\langle card \rangle)$$
$$| \ \langle name1 \rangle, \ldots, \langle nameN \rangle : \langle type \rangle^{\langle card \rangle}$$

$$Cardinality ::= \langle range \rangle \text{ with } \langle frequencyList \rangle; \ldots$$
$$| \ \langle range \rangle \qquad //\text{empty frequencyList}$$

A range has various shorthand forms for the most common cases (continuous range, zero or one, zero or more, one or more, one only).

$$Range ::= \{\langle nos \rangle, \ldots\} \qquad //\text{general}$$
$$| \ \langle nosMin \rangle \ldots \langle nosMax \rangle \qquad //\text{continuous finite range}$$
$$| \ \langle nosMin \rangle \ldots \qquad //\text{continuous infinite range}$$
$$| \ \langle nos \rangle \qquad //\text{a singleton value}$$
$$| \ ? \qquad //\{0,1\}$$
$$| \ * \qquad //0..$$
$$| \ + \qquad //1..$$
$$| \qquad //(\text{blank}) \ 1$$

$$Frequency ::= \langle range \rangle @ \langle proportion \rangle$$

The proportion may be written as a number (for example, 1/2, 0.5, 50%) or as a descriptive phrase (for example, 'most of the time', 'nearly always'); numbers are preferred, but not always known. The superscript form of range tends to be used for simple ranges with no frequencies, whilst the bracketted form is more readable for more complicated expressions.

C.2.5 Predicate

A PREDICATE is a logical expression concerning elements of the model and their properties. The usual logical operators can be used (not, ¬; and, ∧; or, ∨; if, ⇒; iff, ⇔; for all, ∀; exists, ∃).

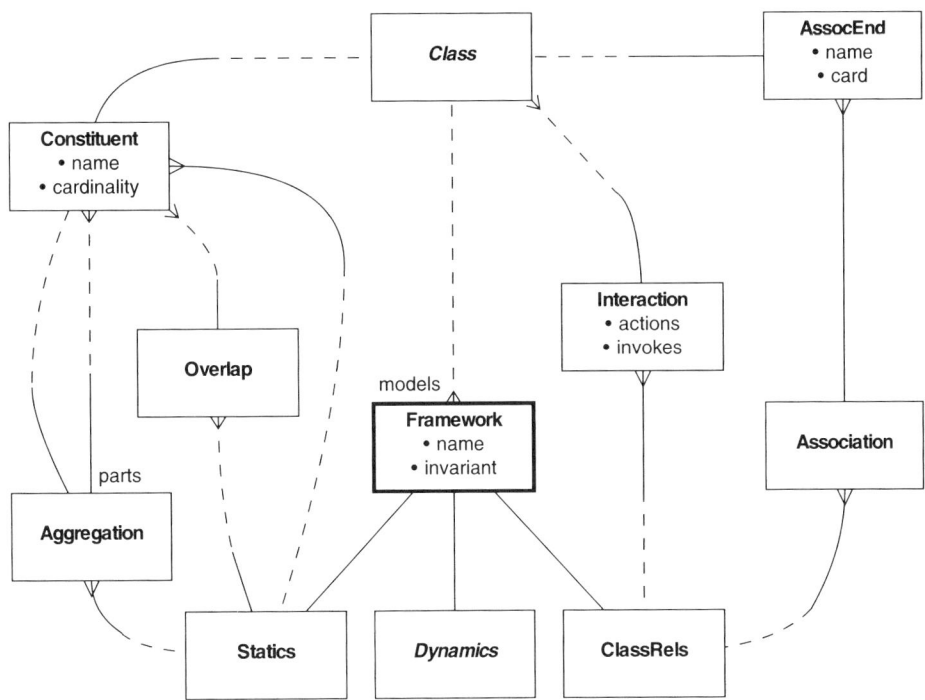

Figure C.2 Framework, and related constructs

C.3 Frameworks

Beluga's *framework* construct is used to describe how its components work together to achieve some coherent behaviour. A framework can be abstracted to a class, in which case the framework can also be thought of as a model of this 'higher level' class in terms of internal components working together to provide the behaviour of the class. A framework's components are defined using collections of stereotypical instances, called *constituent sets*. These constituents are instances of 'lower level' classes, which may themselves be modelled by frameworks. See Figure C.2.

$$\begin{aligned}
Framework \longrightarrow\ & name : \text{C\small LASS}\text{N\small AME} \\
;\ & modelOf : \text{C\small LASS}\text{N\small AME}^? \\
;\ & statics : Statics \\
;\ & classRelationships : ClassRels \\
;\ & dynamics : Dynamics \\
;\ & invariant : \text{P\small REDICATE}
\end{aligned}$$

A framework has a *name*, and, optionally, the name of the class it is a *model of*. The framework definition consists of a description of its *static* components, *class relationships* and *dynamic* behaviour. The framework invariant can be used to talk about properties of collections of constituent sets, for example the topology of their interrelationships (that they are fully connected, say), set relationships between them, etc.

Static semantics

Framework names are in the same name space as class names. The *modelOf* name is optional because a top level framework may not be accompanied by a corresponding class abstraction. The framework will often be the same as the described class's name, but it may need to be different since more than one framework may model the same class. (This is connected to the notion of viewpoints.)

Concrete syntax

The various component parts of a framework are listed under a relevant keyword, and separated by lines to aid readability.

$Framework ::=$

framework ⟨*name*⟩ models	⟨*modelOf*⟩
⟨*statics*⟩	
⟨*classRelationships*⟩	
dynamics	⟨*dynamics*⟩
invariant	⟨*invariant*⟩

C.4 Framework statics

$Statics \longrightarrow constituentSetList : ConstituentSet^*$
$\quad ;\ overlapList : Overlap^*$
$\quad ;\ aggregationList : Aggregation^*$

The *static* part of a framework model is concerned with its *constituent sets*. It describes which constituents occur in the framework, how these constituent sets may share *overlapping* subconstituents, and how constituent parts are *aggregated* into wholes.

Concrete syntax

$Statics ::=$

    constituents ⟨*constituentSetList*⟩;

    . . .

    overlaps ⟨*overlapList*⟩;

    . . .

    aggregations ⟨*aggregationList*⟩;

    . . .

The order of the sections is not significant.

C.4.1 Constituent set

$$ConstituentSet \longrightarrow name : \textsc{ConName}$$
$$;\quad class : \textsc{ClassName}$$
$$;\quad card : Cardinality$$

A constituent set has a *name*, and denotes of a collection of stereotypical instances of a particular *class*. How many instances may occur in a particular constituent set is given by the *cardinality*.

A constituent set's class may itself be modelled by a framework that in turn has constituent sets. These latter constituent sets are termed *subconstituents* of the former.

Advice

If there is more than one instance in a constituent set, the name chosen for the constituent set should be some sort of collective noun.

Concrete syntax

$$ConstituentSet ::= \langle name \rangle : \langle class \rangle \langle card \rangle$$

Example

Consider a constituent set of the framework *Court* that is called *allJuries* and consists of one or more members of class *Jury*. It is written as $allJuries : Jury^+$. If the class *Jury* is modelled by the framework *Jury*, and if the framework *Jury* has a constituent set $clerk : Clerk$ and a constituent set $allJurors : Juror^{12}$, then *clerk* and *allJurors* are both subconstituents of *allJuries*.

C.4.2 Overlap

$$Overlap \longrightarrow nameList : SubConName^+$$

$$SubConName \longrightarrow conNameList : \textsc{ConName}^+$$

An overlap is a list of *subconstituent name* lists. This construct describes the case where different constituent sets have subconstituents in common.

Concrete syntax

$$Overlap ::= \langle nameList \rangle = \ldots$$

$$SubConName ::= \langle conNameList \rangle. \ldots$$

Example

Consider a constituent set called *window*, with one or more instances of class $MVC(window : MVC^+)$. The class MVC may be modelled by a framework that has three constituent sets, a *model*, a *view* and a *controller*, so these are subconstituents of *window*. If we wanted to say that all the *model* subconstituents were in fact the same instance, that they all overlap on *model*, we write

$$window.model$$

Consider a framework *Court* with two constituent sets, $allJuries : Jury^*$ and $allBenches : Bench^+$. Let the class *Jury* be modelled by the framework *Jury*, which has two constituent sets, $allJurors : Juror^{12}$ and $clerk : Clerk$, and the class *Bench* be modelled by the framework *Bench*, which has two constituent sets, $judge : Judge$ and $clerk : Clerk$. To say that the *Bench* and *Jury* frameworks actually share the same clerks when used in the *Court* framework, we write

$$allJuries.clerk = allBenches.clerk$$

C.4.3 Aggregation

$$Aggregation \longrightarrow aggregator : \text{CONNAME}$$
$$; \ partList : \text{CONNAME}^+$$

An aggregation is a collection of constituent sets, one of which, the *aggregator*, is distinguished. Aggregation is one way to combine various *parts* into a whole. The aggregator is a specialised constituent set that can be thought of as gluing the other parts together and being responsible for providing any behaviour of the whole that cannot sensibly be delegated to the parts.

Static semantics

Within a model, the class of the aggregator is used in only one framework (because it is such a highly specialised component), whereas the classes of the parts may be reused in many frameworks. The cardinality of the aggregator's constituent set is one.

Concrete syntax

$$Aggregation ::= \langle aggregator \rangle(\langle partList \rangle, \ldots)$$

Compare a constructor function plus arguments: the aggregator 'constructs' an aggregate entity.

C.5 Framework class relationships

$ClassRels \longrightarrow associationList : Association^*$
$\quad ; \quad interactionList : Interaction^*$

This part of the framework model describes relationships at the class level. It is the part that corresponds most closely to conventional 'entity-relationship diagrams'.

Concrete syntax

$ClassRels ::=$

$\qquad$ associations $\quad \langle associationList \rangle;$
$\qquad \qquad \ldots$
$\qquad$ interactions $\quad \langle interactionList \rangle;$
$\qquad \qquad \ldots$

The order of the sections is not significant.

C.5.1 Associations

$Association \longrightarrow end1, end2 : AssocEnd$

$AssocEnd \longrightarrow name : \textsc{AssocName}$
$\quad ; \quad class : \textsc{ClassName}$
$\quad ; \quad card : Cardinality$

An association between two classes means their instances can be connected in some way, possibly indirectly by a third party. Each *end* of the association has a *name*, these may be the same, and a *cardinality*, indicating how many instances take part in the association. (It is analogous to a relationship in an ERD.)

Static semantics

Each association end must be part of exactly one association. The names of the two ends of an association may be the same. The names of all the association ends of a particular class must be different.

Concrete syntax

$Association ::= \langle end1 \rangle \text{—} \langle end2 \rangle \qquad$ //pretty: em-dash
$\qquad \qquad \ | \ \langle end1 \rangle \text{--} \langle end2 \rangle \qquad$ //ascii: dash, dash

$AssocEnd ::= (\langle name \rangle, \langle class \rangle \langle card \rangle)$

If the two end-names are the same, the Association can be written as follows:

$$Association ::= \langle class \rangle \langle card \rangle \; \overline{\langle name \rangle} \; \langle class \rangle \langle card \rangle$$
$$| \; \langle class \rangle \langle card \rangle \; \text{-}\langle name \rangle\text{-} \; \langle class \rangle \langle card \rangle$$

The pretty form puts the name on the line, the ascii form puts the name 'in' the line.

C.5.2 Interaction

$$Interaction \longrightarrow source, target : ClassName$$
$$; \quad invokes : FeatureName^?$$

In an interaction, an instance of the *source* class invokes a feature in an instance of the *target* class.

Static semantics

If a feature name is given, it must be one of the features defined in the target class.

Concrete syntax

$$Interaction ::= \langle source \rangle \longrightarrow \langle target \rangle.\langle invokes \rangle \qquad //\text{pretty}$$
$$| \; \langle source \rangle\text{-->}\langle target \rangle.\langle invokes \rangle \qquad //\text{ascii}$$

The pretty form uses an arrow, the ascii form uses 'dash, dash, greater'.

C.6 Framework dynamics

The dynamics part of the framework model describes permitted patterns of behaviour. It describes how the constituent sets work together to provide the overall behaviour of the framework. See Figure C.3.

$$Dynamics \longrightarrow \text{NULL}$$
$$| \; \text{STOP}$$
$$| \; FrameworkRef$$
$$| \; Let$$
$$| \; LocalRef$$
$$| \; Participation$$
$$| \; AssociateEvent$$
$$| \; DissociateEvent$$
$$| \; InteractEvent$$
$$| \; Choice$$
$$| \; Composition$$
$$| \; Assertion$$

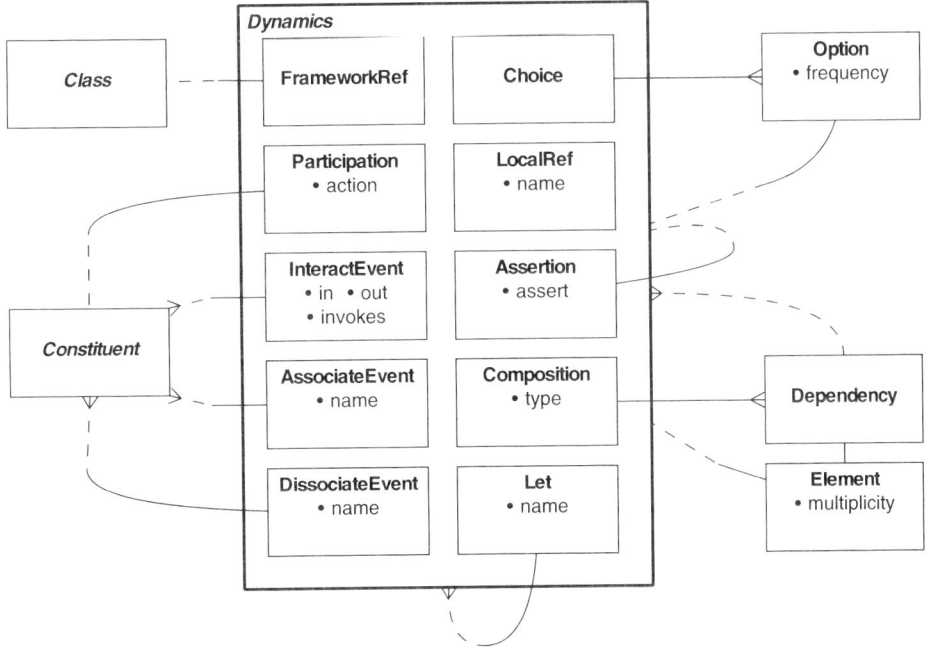

Figure C.3 Framework dynamics

C.6.1 Null behaviour

$Dynamics \longrightarrow \text{NULL}$

One of the kinds of dynamic behaviour is NULL. This is the behavioural equivalent of 'skip' and models 'empty' behaviour. NULL only ever appears in a choice construct; it doesn't mean anything anywhere else.

Concrete syntax

$Dynamics ::= \emptyset$
$\quad\quad\quad\quad | \ \texttt{null} \quad\quad //\text{ascii}$

C.6.2 Stop behaviour

$Dynamics \longrightarrow \text{STOP}$

One of the kinds of dynamic behaviour is STOP. This models a final or 'broken' behaviour. No further behaviour occurs after STOP. If this behaviour is present in a parallel composition, the other branches of the parallel may continue, but no subsequent dependent behaviour occurs.

Concrete syntax

$$Dynamics ::= \dagger$$
$$\mid \text{stop} \qquad //\text{ascii}$$

C.6.3 Framework reference

$$FrameworkRef \longrightarrow ref : \text{ClassName}$$

Framework reference provides a way to *refer* to (the behaviour of) a Framework defined elsewhere.

Concrete syntax

$$FrameworkRef ::= \langle ref \rangle$$

C.6.4 Local definition—'let'

$$Let \longrightarrow name : \text{LocalName}$$
$$;\ localDef, body : Dynamics$$

A local definition allows a piece of dynamic behaviour to be referred to by *name* in the *body*. The scope of the name is the body. Note that *body* may also be a 'let'—there may be multiple local definitions.

Concrete syntax

$$Let ::=$$
$$\quad \text{let } \langle name \rangle == \langle localDef \rangle \text{ in}$$
$$\quad\quad \langle body \rangle \qquad\qquad //\text{`before use' form}$$
$$\mid$$
$$\quad \langle body \rangle$$
$$\quad\quad \text{where } \langle name \rangle == \langle localDef \rangle \qquad //\text{`after use' form}$$

C.6.5 Local reference

$$LocalRef \longrightarrow ref : \text{LocalName}$$

A local reference is a *reference* to a piece of dynamic behaviour that has been declared in a local Let definition.

Static semantics

The definition must be in scope when it is used.

C.6.6 Participation

$Participation \longrightarrow name : \textsc{ConName}$
$\quad\quad\quad\quad\quad\quad ;\ action : Action$

A participation is a piece of behaviour that a *named* constituent set undergoes. It involves one of a certain kind of *action*.

$Action \longrightarrow Initiation$ \quad\quad\quad\quad\quad\quad\quad\quad //creation
$\quad\quad\quad\ \ |\ Termination$
$\quad\quad\quad\ \ |\ Involvement$ \quad\quad\quad\quad\quad\quad\quad //the default
$\quad\quad\quad\ \ |\ Transient$ \quad\quad\quad\quad //for example, a 'signal'
$\quad\quad\quad\ \ |\ Mutation$

There are five kinds of action. Initiation refers to the creation of a constituent set (it engages in no behaviour before initiation), and termination refers to the deletion of a constituent set (it engages in no behaviour after termination). Involvement, the default, is some undifferentiated action of the constituent set. A transient action, for example a 'signal', involves the creation, some behaviour, and immediate termination of the constituent set (it engages in no behaviour before or after the action). In mutation, the constituent set changes state.

$Initiation \quad \longrightarrow statusList : \textsc{StatusName}^*$
$Termination \longrightarrow statusList : \textsc{StatusName}^*$
$Involvement \longrightarrow statusList : \textsc{StatusName}^*$
$Transient \quad \longrightarrow statusList : \textsc{StatusName}^*$
$Mutation \quad \longrightarrow changeList : Change^+$
$Change \quad\quad \longrightarrow before, after : \textsc{StatusName}$

All the actions refer to sets of *statuses*. These are the permitted statuses of the constituent set instances after initiation, during involvement and transience, or before termination. A mutation consists of a set of status pairs, corresponding to the permitted relationships between statuses *before* and *after* the mutation.

Concrete syntax

$Participation ::= \langle action \rangle \langle name \rangle$

$Action ::= \text{init}\{\langle statusList \rangle, \ldots\}$ \quad\quad\quad //Initiation
$\quad\quad\ \ |\ \{\langle statusList \rangle, \ldots\}$ \quad\quad\quad\quad\quad //Involvement
$\quad\quad\ \ |\ \text{mut}\{\langle changeList \rangle, \ldots\}$ \quad\quad\quad //Mutation
$\quad\quad\ \ |\ \text{trans}\{\langle statusList \rangle, \ldots\}$ \quad\quad\quad //Transient
$\quad\quad\ \ |\ \text{term}\{\langle statusList \rangle, \ldots\}$ \quad\quad\quad //Termination

If the *statusList* is empty, the braces are omitted.

$Change ::= \langle before \rangle \Rightarrow \langle after \rangle$ \quad\quad //pretty: double arrow
$\quad\quad\ \ |\ \langle before \rangle \text{=>} \langle after \rangle$ \quad\quad //ascii: equals, greater

C.6.7 Associate event

$$AssociateEvent \longrightarrow name : AssocName$$
$$;\ party1, party2 : ConName$$

An associate event behaviour creates an instance of a *named* association between two given *parties*.

Static semantics

The two parties' classes must be related by an Association, and the name must be one of the end-names of that Association. The parties must not initially be associated.

Concrete syntax

$$AssociateEvent ::= \langle party1 \rangle \text{—} \langle party2 \rangle : \langle name \rangle \qquad //\text{pretty: em-dash}$$
$$|\ \langle party1 \rangle \text{--} \langle party2 \rangle : \langle name \rangle \qquad //\text{ascii: dash, dash}$$

Note the 'instance : type' syntax (compare ConstituentSet declarations).

C.6.8 Dissociate event

$$DissociateEvent \longrightarrow name : \textsc{AssocName}$$
$$;\ party1, party2 : \textsc{ConName}$$

A dissociate event behaviour dissolves an instance of a named association between two given parties

Static semantics

The two parties' classes must be related by an Association, and the name must be one of the end-names of that Association. The parties must initially be associated.

Concrete syntax

$$DissociateEvent ::= \langle party1 \rangle \text{-x-} \langle party2 \rangle : \langle name \rangle$$

C.6.9 Interact event

$$InteractEvent \longrightarrow source : \textsc{ConName}^?$$
$$;\ target : \textsc{ConName}$$
$$;\ invokes : \textsc{FeatureName}^?$$
$$;\ ins, outs : ActualParam^*$$

In an interact event behaviour, instances of the *source* constituent set invoke a feature on instances of the *target* constituent set, supplying *input* parameters and receiving the *outputs*. The source is optional to permit 'anonymous' events, or events originating from outside the model, to occur. The feature name is optional to allow 'early' models of behaviour, before the relevant class has been fully defined, to be syntactically correct.

$$ActualParam \longrightarrow ClassAParam$$
$$| \quad SimpleAParam$$

$$ClassAParam \longrightarrow name : \textsc{ConName} \qquad \text{//a ConstituentSet}$$
$$SimpleAParam \longrightarrow name : \textsc{Name} \qquad \text{//a 'value identifier'}$$

Static semantics

The two parties' classes must be related by an Interaction on the relevant feature. If a feature name is given, the actual parameters must be type-consistent with the feature's signature.

Concrete syntax

$$InteractEvent ::= \langle source \rangle \longrightarrow \langle target \rangle . \langle invokes \rangle (\langle ins \rangle, \ldots) : (\langle outs \rangle, \ldots)$$

The ascii form uses '-->' (dash, dash, greater) instead of the arrow.

Static semantics

It is possible to have in/out parameters without a feature name.

C.6.10 Choice

$$Choice \longrightarrow optionList : Option^+ \qquad \text{//at least two}$$

$$Option \longrightarrow body : Dynamics$$
$$; \quad frequency : Frequency^?$$

A choice behaviour provides a choice between two or more behaviours. The various *options* may be guarded with assertions (see later), in which case an assertion must be true for its behaviour to be chosen. If more than one assertion is true, or an option is unguarded, then one of the behaviours is chosen non-deterministically, with a probability weighted by the optional *frequency*. (Having one branch of a choice being a low frequency STOP allows the non-functional property of failure rate to be expressed.) If no frequencies are given, all the branches have equal weight.

If all the options are guarded with false assertions, the behaviour 'waits' for one assertion to become true (for example, a timeout, or a value changed by a parallel behaviour). If no assertion can ever become true, the behaviour is equivalent to STOP.

Static semantics

There must be at least two options in a choice. (One option would be equivalent to that behaviour on its own.)

Concrete syntax

$$Choice ::= (\langle optionList \rangle \mid \ldots)$$

$$Option ::= \langle body \rangle @ \langle frequency \rangle$$

C.6.11 Composition

$$Composition \longrightarrow dependencyList : Dependency^+$$
$$;\quad type : CompositionType$$

A composition composes behaviours either sequentially (where behaviours whose order is not determined nevertheless cannot be interleaved; one must finish before the next can start) or in parallel. The syntax permits a general graph structure of dependencies.

$$CompositionType \longrightarrow \text{PARALLEL}$$
$$\mid \text{SEQUENTIAL}$$

$$Dependency \longrightarrow precursorList : Dynamics^*$$
$$;\quad dependent : Element$$

$$Element \longrightarrow elem : Dynamics$$
$$;\quad multiplicity : Multiplicity$$

$$Multiplicity \longrightarrow Cardinality$$
$$\mid Mapping$$

$$Mapping \longrightarrow element, set : \text{CONNAME}$$

Concrete syntax

$$Composition ::= [\langle dependencyList, \ldots \rangle] \qquad\qquad //\text{parallel}$$
$$\mid \; [\![\langle dependencyList, \ldots \rangle]\!] \qquad //\text{pretty: sequential}$$
$$\mid \; [[\langle dependencyList, \ldots \rangle]] \qquad //\text{ascii: sequential}$$

Adjacent square brackets are always treated as a sequential composition construct; nesting parallel compositions is semantically redundant.

$$Dependency ::= \langle precursorList \rangle \& \ldots \gg \langle dependent \rangle$$

Components in the *precursorList* are normally referred to by name (using a local reference). In the special case where the dependency graph forms a linear succession (where the precursor of component $n+1$ is the single component n), the composition is written as follows, without brackets:

$Composition ::= \langle dependent1 \rangle \gg \ldots$

$Element \quad ::= \langle elem \rangle^{\langle multiplicity \rangle}$

$Mapping \quad ::= \langle element \rangle / \langle set \rangle$

Example

Consider the constituent set $bb : Book^*$, zero or more instances of a *Book*. If the behaviour *DoBook* refers to a constituent set b, then $[DoBook^{b/bb}]$ does *DoBook*, in parallel, for each element of bb bound to b, whereas $[\![DoBook^{b/bb}]\!]$ does one book at a time, but in no specified order.

C.6.12 Assertion

$Assertion \longrightarrow assert : \text{PREDICATE}$
$\qquad\qquad\quad ; \;\; body : Dynamics$

In an assertion behaviour, the *assertion* is true when the *body* occurs. Assertions can either be about the features of constituent set instances, or about the temporal status of other bits of behaviour. Dependencies and sequentiality (in sequential composition) are special cases of the latter. Interrupts and timeouts can also be handled in this manner.

Concrete syntax

$Assertion ::= \langle assert \rangle ? \langle body \rangle$

Example

$A \gg B = [A, (\text{finished } A)?B]$

$[\![A, B]\!] = [(\text{not } ongoingB)?A, (\text{not } ongoingA)?B]$

C.7 Diagrammatic concrete syntax

C.7.1 Class

A class is drawn as a rectangular box with the class name inside (Figure C.4).

Feature names are grouped together by facet name within a class box (Figure C.5). Feature signatures and meanings can be included if desired. If these clutter the diagram, an accompanying textual form is preferred.

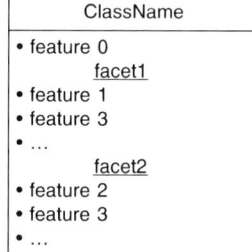

Figure C.4 Class

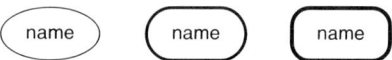

Figure C.5 Class showing features and facets

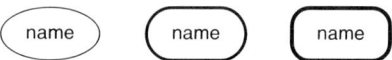

Figure C.6 Constituent set

C.7.2 Frameworks

A framework is drawn as a rectangular box containing the framework name and some elements of diagrammatic syntax (for example, classes, class relationships, constituent sets, timelines). The outermost box of a diagram (corresponding to the diagram border) can be omitted.

Where the name of the framework is different from the name of the class it models, this should be stated in the name component:

⟨name⟩ models ⟨modelOf⟩

C.7.3 Constituent set

A constituent set is drawn as a round-cornered box or an ellipse—the roundness distinguishes it from a class box—with the constituent set name inside (Figure C.6).

The class and cardinality of a constituent set can be indicated in two ways: by using a long-dashed line to link the constituent set box with a class box (Figure C.7) or by putting the full textual form inside the box (Figure C.8). The name of a constituent set may be null. Note that there may be more than one constituent set with the same class (Figure C.9). Sometimes, one constituent set is a subset of

C.7 Diagrammatic concrete syntax

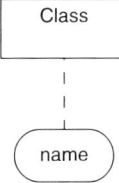

Figure C.7 Cardinality

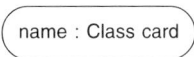

Figure C.8 Cardinality

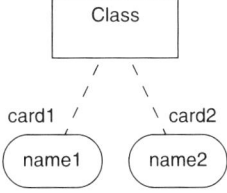

Figure C.9 Cardinality

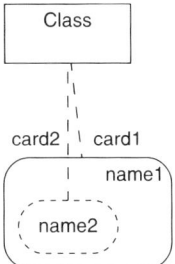

Figure C.10 Cardinality, $card1 > card2$

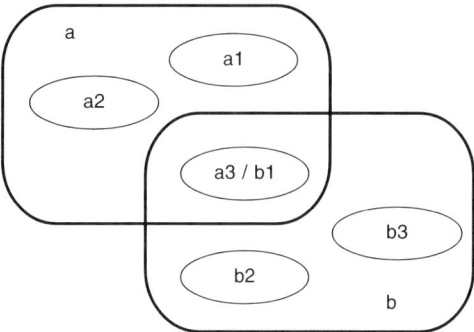

Figure C.11 Overlap. The textual form is $a.a3 = b.b1$

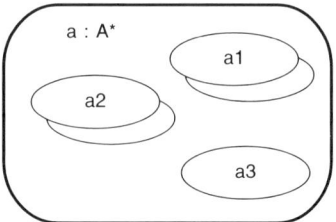

Figure C.12 Overlap. The textual form is $a.a3$

another; this is indicated by nesting a dotted constituent set box (Figure C.10).

Where a framework models a class of the same name, a framework box, rather than a class box, may be used.

C.7.4 Overlap

There are two special diagrammatic forms for overlap, corresponding to the two common cases.

Where the cardinality of a and b is 1, an overlap is drawn by superimposing shared subconstituents (Figure C.11).

A different form is used in the case of an overlap involving a common subconstituent of a constituent set. If constituent set a consists of many instances of A, each of which consists of one $a1$, one $a2$ and one $a3$, where the instances of A share the same $a3$ constituent set, it can be drawn as in Figure C.12.

C.7.5 Aggregation

An aggregation is shown using a 'diamond' connector (Figure C.13). Where all the constituent sets of a given class in a framework are aggregated, an aggregation

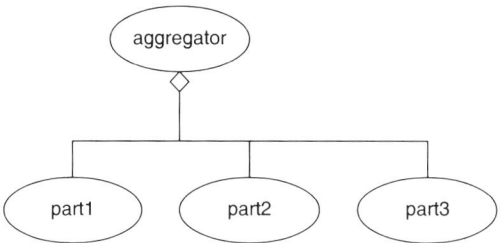

Figure C.13 Aggregation between constituent sets

Figure C.14 Aggregation between classes

Figure C.15 Association

structure can be drawn between class boxes (Figure C.14). This allows aggregations to be drawn at the class level, along with associations and interactions. See Chapter 12 for more discussion.

C.7.6 Association and interaction

An association is drawn as a straight line or as a curved line between class boxes (Figure C.15). A common name may be drawn in the middle of the line (Figure C.16).

Multiple cardinality may be indicated by decorating line-ends. We use a standard entity-relationship convention, where dashed lines represent optionality, and 'crows-feet' represent 'one or more'. Figures C.17–C.20 may be read from left-to-right, to find b's cardinality. They may be read similarly from right-to-left, to

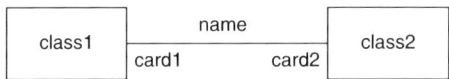

Figure C.16 Association, with the same name at each end

Figure C.17 *a* may optionally be associated with one *b*. So *b* has cardinality $\{0, 1\}$ with textual form ?

Figure C.18 *a* is associated with one *b*. So *b* has cardinality 1 with textual form (blank)

Figure C.19 *a* may optionally be associated with one or more *b*s. So *b* has cardinality 0.. with textual form *

Figure C.20 *a* is associated with one or more *b*s. So *b* has cardinality 1.. with textual form +

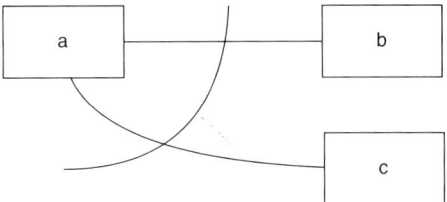

Figure C.21 Exclusion

Figure C.22 Interaction

give *a*'s cardinality.

An exclusion is a particular kind of framework invariant that says each instance of *class*1 is associated with instances of *class*2 or of *class*3, but not both, at any one time. It is shown by a curved line cutting across the relevent association lines (Figure C.21).

An interaction is drawn with a straight or curved arrowed line between class boxes (Figure C.22). The arrowhead is not at the end of the line in order to avoid confusion with the generalisation arrow (see later). An interaction arrow can go to a specific feature or facet of a class, where these are given explicitly (Figure C.23).

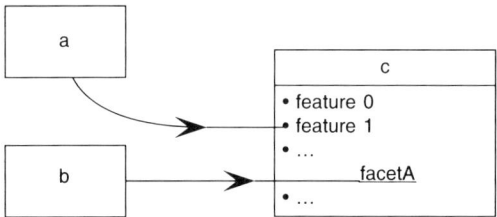

Figure C.23 Interaction, to a feature and facet

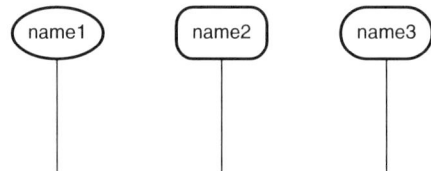

Figure C.24 Timelines

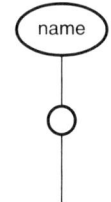

Figure C.25 Participation

C.7.7 Timelines

Timelines are used to indicate framework dynamics. Timelines are drawn from constituent set boxes, with time progressing away from the boxes (Figure C.24).

All timelines within a framework should be drawn with the same orientation—all downwards or all left-to-right.

C.7.8 Participation and action

A participation is drawn as an action icon on the timeline for the named constituent set (Figure C.25). Figure C.26 shows the icons for each kind of action. Statuses

C.7 *Diagrammatic concrete syntax*   275

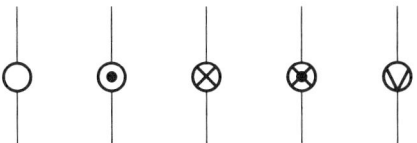

Figure C.26 Action icons

Figure C.27 Status, in involvement and mutation

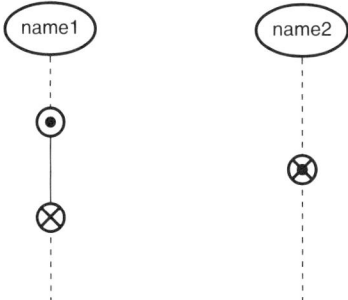

Figure C.28 Lifetime

are written alongside the action icons (Figure C.27). Before and after statuses in a mutation are separated by a line.

Timelines are dotted before initiation, after termination, and either side of transience (Figure C.28). The solid portion of the line thus indicates the extent of the constituent set's 'lifetime'.

C.7.9 Events

Note the similarity between events and class relationships.

An associate event is drawn as a line between participations (Figure C.29). If the constituent sets involved are of different classes, there is no confusion about the 'direction' of the relationship; if the ends of the association have different names, either name can be used. However, if an association is between instances of the

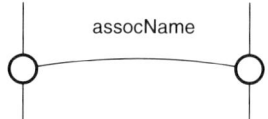

Figure C.29 Associate event

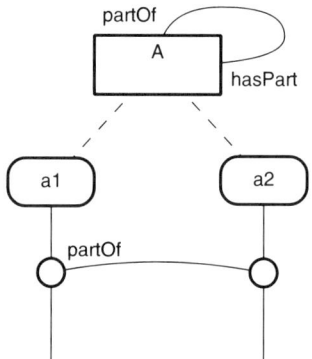

Figure C.30 Named Associate event

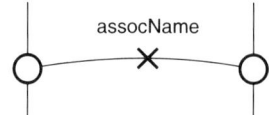

Figure C.31 Dissociate event

same class, the 'direction' may be important, and different names are used for the ends of the association (for example, *hasPart* and *partOf* in Figure C.30). One of the names must be written at the appropriate end of the line.

A dissociate event is drawn with a cross on the line (Figure C.31).

An interact event has an arrow on the line, with (optionally) a feature name at the head-end (Figure C.32). If a third constituent set is a parameter to the interact event, it can be connected to the arrowhead by a dotted line (Figure C.33).

C.7.10 References

References to named frameworks are shown as framework boxes positioned on the timelines (Figure C.34). Local references are usually expanded.

C.7 *Diagrammatic concrete syntax* 277

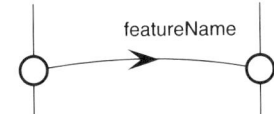

Figure C.32 Interact event

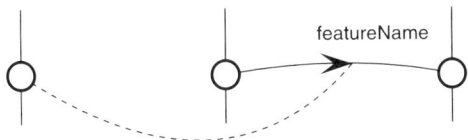

Figure C.33 Interact event, with parameter

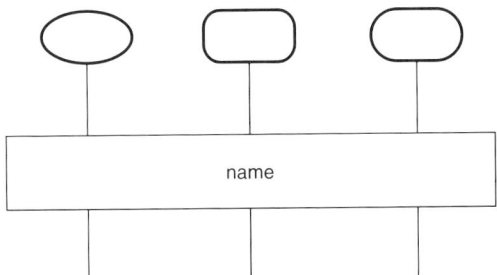

Figure C.34 Reference to a named framework

C.7.11 Composition

The network of dependencies between elements of composition can be drawn using double-headed arrows (compare the textual form for dependencies). The whole composition is put in a box—a single-sided box if it is a parallel composition (Figure C.35), and a double-sided box if it is a sequential composition (Figure C.36), by analogy to the single and double square brackets in the textual form.

Elements can be either framework references (to a named 'event' framework—see above), or as 'subdiagrams'. In the subdiagram case, multiplicity of elements is indicated at the top-right corner of the boxes (Figure C.37). If the subdiagrams share some or all of their constituent sets, the timelines can be linked, with branching to accommodate independent behaviours (Figure C.38). Where the dependencies

278   *Appendix C   Beluga—the Behavioural modelling language*

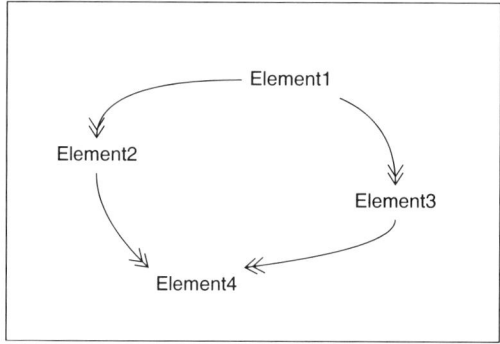

Figure C.35 Parallel composition

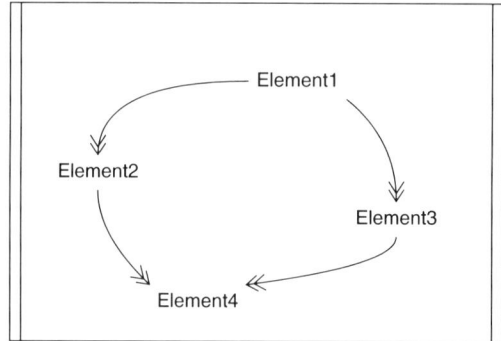

Figure C.36 Sequential composition

are linear, timelines do not need to branch and can thus be straight.

These conventions can apply recursively to subboxes, subsubboxes, and so on. The elements of lowest level boxes will be events (see above) or just participations (if the dynamics are indeterminate). Positioning of events and participations on timelines indicates temporal dependency.

C.7.12   Choice and assertion

A choice construct is boxed, with options separated by horizontal dashed lines (Figure C.39). Where the different options involve common constituent sets, the timelines can be continuous. Guard assertions on options (if any) appear at the top-left of the option boxes. Note that assertions can also be used in compositions, and follow the same convention. If an 'else NULL' option is required, it is represented by a horizontal line below the last option, followed by the NULL symbol or keyword.

C.8 Diagrammatic concrete syntax 279

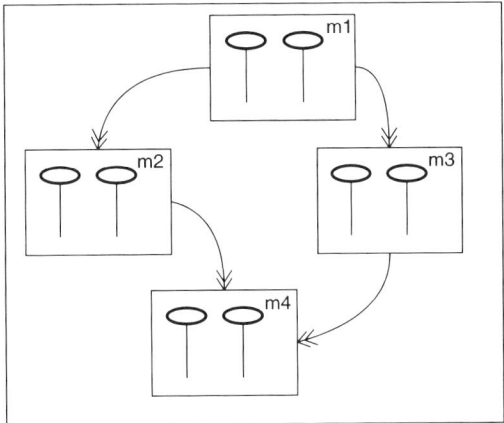

Figure C.37 Subdiagrams, with multiplicities

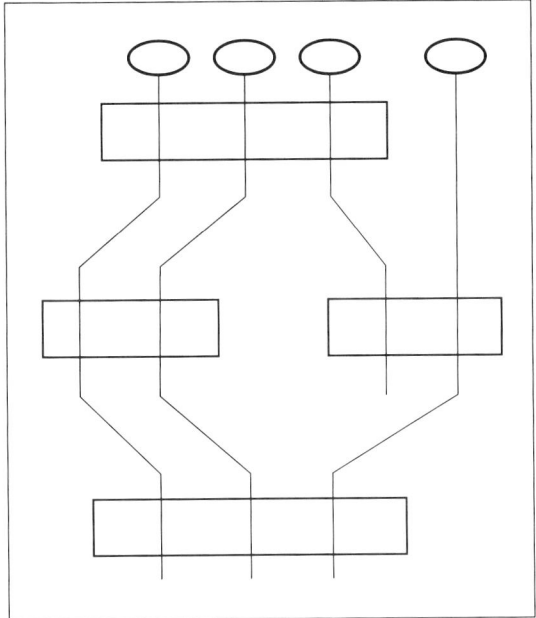

Figure C.38 Linked branching timelines

280   Appendix C   Beluga—the Behavioural modelling language

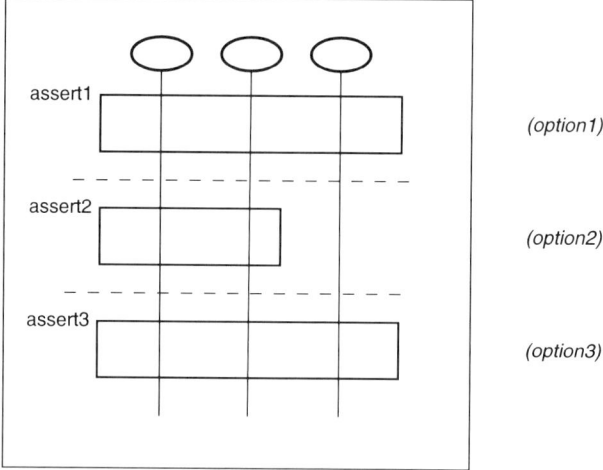

Figure C.39 A three option choice

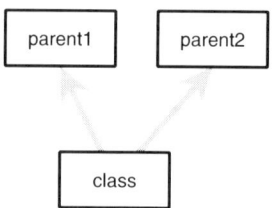

Figure C.40 Generalisation (inheritance)

C.8 Generalisation and abstraction

Generalisation relationships (inheritance) between classes are shown as grey arrows pointing to the parent(s) (Figure C.40). Generalisation arrows are drawn thick, shaded and usually straight in order to avoid confusion with interaction arrows (see earlier).

A framework (a model of a class) uses classes at lower levels of abstraction. This happens in the following two ways:

- Framework static constituent sets are declared to be of particular classes.
- Framework dynamics references other frameworks defined elsewhere.

A framework's structural decomposition, involving constituent set classes occurring in the *fwk.statics.constituentSet.class* part of the model, is shown using a 'pyramid' symbol (Figure C.41). A framework's 'temporal' decomposition, involving frameworks and classes occurring in the *fwk.dynamics.frameworkRef.ref* part of

C.8 Generalisation and abstraction

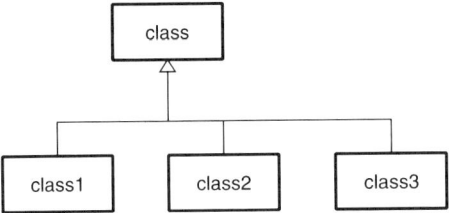

Figure C.41 Framework, structural decomposition

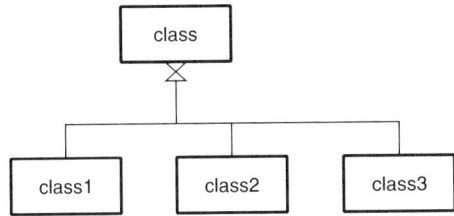

Figure C.42 Framework, temporal decomposition

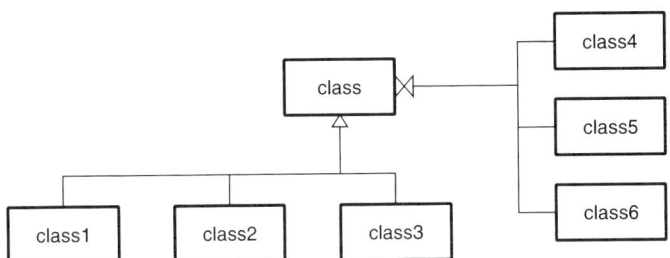

Figure C.43 Framework, structural and temporal decomposition

the model, is shown using an 'hourglass' symbol (Figure C.42). The two modes of use can be shown together. A useful convention is to lay out constituent set use horizontally and temporal structure vertically (Figure C.43).

Constituent set classes may themselves be modelled by frameworks; temporal frameworks may themselves have constituent sets and subframeworks. So there may be multiple levels of abstraction in both the structural and temporal dimensions.

If a class is modelled by more than one framework, the use-structures can be labelled (Figure C.44).

Generalisation and abstraction relationships may be shown in the same diagram

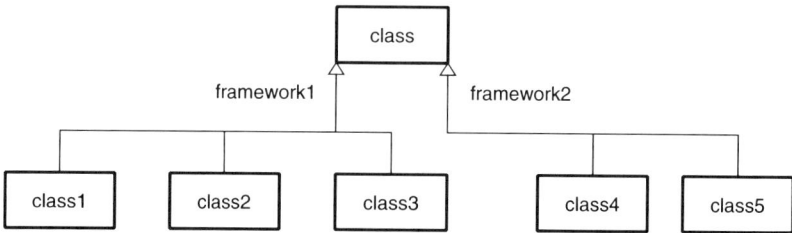

Figure C.44 Framework, multiple models

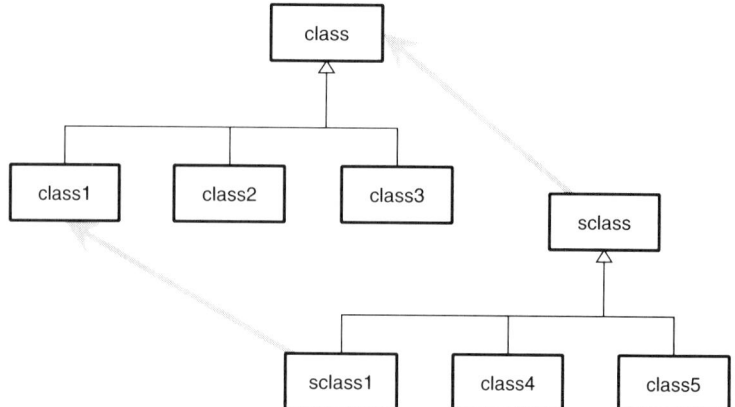

Figure C.45 Generalisation and abstraction

(Figure C.45). This can be useful in illustrating how specialisation propagates between different levels of abstraction.

Appendix D

Glossary

Key

bold primary term

[Grampus] Purposive modelling related term

[Beluga] Behavioural modelling related term

abstraction, level of The resolution at which the world is described. [Grampus] The level at which a purposive entity is expressed as a formation or role. [Beluga] The level at which a behavioural entity is expressed as a framework or class.

aggregation [Beluga] The binding together of various objects by another object. Such an aggregator object exists only to construct a more complex entity (that is, an entity at a higher level of abstraction). Aggregations are commonly expressed between classes, rather than objects.

Ahab A Help in Analysing Balderdash: ORCA's technique for structuring and clarifying raw textual information. (For when you don't have a leg to stand on.)

analysis The activity that aims to provide a context, a rationale and a specification for change to the world.

analysis boundary The definition of which parts of the client's world are open to investigation by an analyst.

assertion [Beluga, dynamics] A statement of conditions that must hold when an episode starts (a pre-assertion), or ends (a post-assertion). These conditions may concern the state of participants in the episode, or the temporal status of other episodes (for example: completed, ongoing, not started.)

associate event [Beluga, dynamics] An event that creates instances of an association between the objects in two constituent sets.

association [Beluga] A kind of pairing between instances of two classes (possibly the same class). Associations are directional and can be named from either end. Each end of an association has a cardinality. Associations are expressed as relationships between classes.

attribute [Beluga] A property of an object explicitly defined by that object's class.

behavioural entity Something in the real world represented as an object or as an episode in a behavioural (Beluga) model.

Beluga BEhavioural Language Underpinning Generic Analysis: ORCA's behavioural modelling language. (So called because 'It'll be all white on the night'.)

cardinality [Beluga] A description of the number of elements in a set, consisting of a range of values and, optionally, the typical frequencies of subranges or individual values.

choice [Beluga, dynamics] An episode that may take different forms or may optionally happen (the latter is expressed as a choice between a substantive event and the null event). In a multi-way choice, assertions may be used to indicate the conditions that hold when particular options happen.

class [Beluga] A characterisation of objects in terms of attributes and operations. A class can be modelled by a framework (at a lower level of abstraction).

client The 'owner' of the part of the world that is the focus of analysis. The client is assumed to be the person who instigates the analysis and who is in charge of implementing change.

cluster [Grampus] A collection of roles linked by co-operations.

component [Grampus] *see* role.

composition [Beluga, dynamics] The combination of multiple frameworks into a more complex framework. Parallel composition is the general case, typically used for combining 'processes'. Sequential composition is typically used for multi-episode 'transactions'. There may be temporal dependencies between elements of a composition. Elements of a composition have cardinalities.

configuration management The activity of managing change to the products of analysis or design, and aggregation of components into deliverables.

constituent set (constituent) [Beluga] In a framework, a collection of stereotypical objects (instances of a single class). A constituent set may be named and has a cardinality (which may be one). The contents of a constituent set may change over time as objects are initiated and terminated. A constituent set may be a participant in one or more events.

co-operation [Grampus] The satisfaction of a reliance by one or more guarantees.

delegation [Grampus] The arrangement whereby a guarantee at one level of a formation is fulfilled by one or more guarantees of roles at the next lower level of formation.

dependency [Beluga, dynamics] There may be dependencies between elements (subframeworks) in a composition. An element can depend on one or more precursor elements, which must have completed before the dependent event can start. A linear sequence of elements is a special case of dependencies within a composition.

design The activity of developing and installing Information Technology, and procedures relating to it.

development The design, construction and installation of facilities as part of a wider programme of system change.

dissociate event [Beluga, dynamics] An event that removes instances of an association between objects in two constituent sets.

domain In analysis: a subject area with which a particular part of the world deals (for example: banking, air traffic control, telephony). In design: areas of functionality within an implemented IT system (for example: user Interface, application domain, communications).

dynamic behaviour That aspect of a system concerned with patterns of behaviour over time (as opposed to static structure).

episode [Beluga] A piece of behaviour, of finite duration, that is characterised by a framework. An episode can combine simpler episodes, and ultimately events. An episode is regarded as having a start time, a finish time and a duration.

event [Beluga] A piece of behaviour that is characterised by an atomic construct within a given framework (for example: associate event, dissociate event, interact event).

extrinsic [Grampus] A guarantee or reliance attributed to a role by some other role (as opposed to intrinsic).

facet [Beluga] A grouping of the features defined by a class. Typically, such a facet groups the features involved in a particular aspect of behaviour.

feature [Beluga] A general term for the attributes and operations defined by a class. A feature can have a signature, which defines the type of the attribute or the types of the parameters and result of the operation.

formation [Grampus] The combination of two or more roles into a role at a higher level of abstraction. Reliances and guarantees at the two levels may be related by delegation and promotion.

framework [Beluga] A characterisation of a system in terms of the classes of objects involved and the possible histories of objects and their interactions. A framework describes one or more constituent sets of objects, their classes, the static structure, dynamic behaviour and invariant properties. A framework can model a class (that is, characterising a behavioural entity at a lower level of abstraction).

generalisation [Beluga] A class is a generalisation of another class if the former class characterises a superset of the objects characterised by the latter class. The opposite of generalisation is specialisation.

generic model A (Beluga) model that characterises a variety of different systems.

Grampus Guarantee & Rely Approach to Modelling Purpose in Systems: ORCA's purposive modelling language.

guarantee [Grampus] A description of a service that a role guarantees is provided.

history [Beluga] The observable behaviour of a system over some period. A compound entity characterised by a framework.

information gathering The activity of eliciting and recording information by direct observation, or from information sources (for example: the client, employees, potential users, documents).

Information Technology (IT) A term covering computer hardware and software, communications links and input/output devices.

initiation [Beluga, dynamics] The first participation in which a constituent set is involved.

interact event [Beluga, dynamics] An event in which the objects in a constituent set (the source) change the state of some objects in another constituent set (the target). The state-change that takes place may be defined to be a change in status, a change in associations, or the result of an operation defined by the target class.

interaction [Beluga] The relationship between two classes whereby instances of the source class may change the state of instances of the target class. An interaction may be defined as invoking an operation defined by the target class.

intrinsic [Grampus] A guarantee or reliance attributed by a role to itself (as opposed to extrinsic).

invariant [Beluga] In a framework, a statement of properties that hold throughout a history (for example: fixed structural relationships, timing constraints).

method A disciplined approach, consisting of a tailorable process, one or more modelling languages, and heuristics.

Narwhal Not A Really Wonderfully Helpful Analysis Language: one of the many whales that is not an ORCA modelling language.

New World The world after the implementation of some programme of change and development (as opposed to Old World).

non-functional requirements A general term for requirements concerning security, reliability, capacity, timing, and so on. ORCA treats all these properties as aspects of behaviour.

object [Beluga] A behavioural entity that is characterised by a class; a unitary (rather than compound) entity.

Old World The world before the implementation of some programme of change and development (as opposed to New World).

operation [Beluga] The capability of changing the state of an object in some particular way by acting directly on that object. The operations applicable to an object are defined by its class.

parameter [Beluga] The parameters of an operation are objects or values that must be passed to the target object in order to invoke the operation.

participant [Beluga, dynamics] The constituent set referenced by a participation.

participation [Beluga, dynamics] The involvement of a constituent set in an event.

pathology An identified set of factors responsible for observed problems with a system.

prescription An identified set of changes to a system that will remedy a particular pathology.

process [Beluga] A piece of behaviour, of indefinite duration, that is characterised by a framework. A process can combine simpler processes or repeating episodes.

process design The activity of determining and documenting the way in which analysis or development is to be carried out.

promotion [Grampus] The arrangement whereby a reliance at one level of a formation arises from one or more reliances of roles at the next lower level of formation.

purposive entity Something in the world represented as a role in a purposive (Grampus) model.

reliance [Grampus] A description of a service that a role requires be provided.

rich picture An informal pictorial representation of a client's 'world'. Typically produced prior to modelling as part of a scoping and familiarisation activity.

role [Grampus] A description of a purposive entity in terms of its reliances and guarantees, and co-operations with other purposive entities.

service [Grampus] An aspect of behaviour which a co-operation concerns. Guarantees and reliances are views of service provision.

specialisation [Beluga] A class is a specialisation of another class if the former class characterises a subset of the objects characterised by the latter class. The opposite of specialisation is generalisation.

state [Beluga] The state of an object consists of values for its attributes, together with its associations with other objects (and relevant properties of associated objects).

status [Beluga, dynamics] The states of objects in a constituent set can be distinguished by simple labels denoting different statuses. (For example, a library book can be 'in' or 'out'.)

static semantics The rules for well-formed models in a particular modelling language (for example: use of names, structural constraints).

static structure That aspect of a system concerned with its constituent objects, their classes, and their organisation (as opposed to dynamic behaviour).

subclass [Beluga] If a class is a specialisation of another class, the former is a subclass of the latter.

superclass [Beluga] If a class is a generalisation of another class, the former is a superclass of the latter.

syntax A definition of the rules for expressing information using a modelling language. The abstract syntax of a modelling language defines the underlying constructs, while (one or more) concrete syntaxes provide ways of representing these constructs in textual or diagrammatic form.

system A part of the world that is delimited and coherent, both purposively and behaviourally.

termination [Beluga, dynamics] The final participation in which a constituent set is involved.

timeline [Beluga] In the diagrammatic representation of a framework, a timeline drawn from a constituent set box can be used to order the events in which that constituent set is a participant.

validation The activity that aims to ensure the accuracy and adequacy of models produced within analysis.

viewpoint [Grampus] The description of a role in terms of its intrinsic guarantees and reliances, together with any extrinsic guarantees and reliances associated with these (that is, attributed to other roles).

Appendix E

Bibliography

[Barden *et al.* 1994]
:   Rosalind Barden, Susan Stepney, and David Cooper. *Z in Practice*. BCS Practitioners Series. Prentice Hall, 1994.

[Booch 1991]
:   Grady Booch. *Object Oriented Design with Applications*. Benjamin-Cummings, 1991.

[Checkland & Scholes 1991]
:   Peter Checkland and Jim Scholes. *Soft Systems Methodology in Action*. Wiley, 1991.

[Checkland 1981]
:   Peter Checkland. *Systems Thinking, Systems Practice*. Wiley, 1981.

[DeMarco 1978]
:   Tom DeMarco. *Structured Analysis and System Specification*. Yourdon Press, Prentice-Hall, 1978.

[Gardner 1990]
:   Martin Gardner. *Gardner's Whys and Wherefores*. Oxford University Press, 1990.

[Goldstein & Alger 1992]
:   Neal Goldstein and Jeff Alger. *Developing Object-Oriented Software for the Macintosh: analysis, design and programming*. Addison-Wesley, 1992.

[Harel 1987]
:   David Harel. Statecharts: A visual formalism for complex systems. *Science of Computer Programming*, 8(3):231–274, 1987.

[Jacobson 1992]
:   Ivar Jacobson. *Object-Oriented Software Engineering: a Use Case driven approach*. Addison-Wesley, 1992.

[Martin & Odell 1992]
: James Martin and James J. Odell. *Object-Oriented Analysis and Design.* Prentice Hall, 1992.

[Meyer 1990]
: Bertrand Meyer. *Introduction to the Theory of Programming Languages.* Prentice Hall, 1990.

[Patching 1990]
: David Patching. *Practical Soft Systems Analysis.* Pitman, 1990.

[Rumbaugh *et al.* 1991]
: James Rumbaugh, Michael Blaha, William Premerlani, Frederick Eddy, and William Lorensen. *Object-Oriented Modeling and Design.* Prentice Hall, 1991.

[Spivey 1992]
: J. Michael Spivey. *The Z Notation: a Reference Manual.* Prentice Hall, 2nd edition, 1992.

[Tufte 1983]
: Edward R. Tufte. *The Visual Display of Quantitative Information.* Graphics Press, 1983.

[Tufte 1990]
: Edward R. Tufte. *Envisioning Information.* Graphics Press, 1990.

[Wirfs-Brock *et al.* 1990]
: Rebecca J. Wirfs-Brock, Brian Wilkerson, and Lauren Wiener. *Designing Object-Oriented Software.* Prentice Hall, 1990.

Appendix F
Index

abstract class, 127
abstract syntax, 186, 237
abstraction, 6, 30, 140, 152, 283
abstraction map, 149
access to information, 50
activity area, 60
adequate, 5
aggregation, 144, 145, 195, 258, 270, 283
Ahab, 283
alfalfa, 117
analysis, 3, 283
analysis boundary, 46, 283
analysis project, 50
animation, 222, 228
architecture, 111
assertion, 267, 278, 283
associate event, 121, 129, 264, 275, 283
association, 23, 121, 129, 259, 271, 284
attribute, 16, 125, 284
audit trail, 228

balderdash, 208
Basic Process, 5, 25, 38, 52, 157
Batching System, 205
behaviour, 6
behavioural entity, 116, 284

behavioural model, 16, 26
Beluga, 284
Birfami, 117
brocade weft, 44
bundle of prescriptions, 97
Business Process Re-engineering, 169, 189

cam shaft, 73
capacity requirements, 224
cardinality, 284
choice, 284
choice behaviour, 265, 278
class, 16, 146, 251, 267, 284
class relationship, 147
classification, 152
client, 284
client participation, 222
client–server, 7
cluster, 245, 284
co-operation, 12, 26, 55, 245
co-ordination problem, 165
completeness, 150
component, 284
composition, 136, 266, 277, 284
concrete syntax, 186, 237
concurrency, 136
configuration management, 226, 228, 284

constituent set, 19, 135, 257, 268, 284
contractual agreement, 50
control box, 73
control system, 74, 204
cooperation, 284
critical success factor, 173, 233
crows-foot, 16, 126, 271

delegation, 30, 57, 248, 284
deliverables, 221
dependency, 124, 138, 284
design, 285
development, 285
development constraint, 225
development objectives, 49, 89
dissociate event, 121, 129, 264, 276, 285
domain, 285
dynamic behaviour, 6, 147, 285
dynamics, 16

electromagnet, 43, 73
environment, 29
episode, 18, 136, 285
essential, 5
event, 121, 285
exclusion, 273
extensibility, 225
extrinsic, 27, 60, 163, 176, 244, 285

facet, 126, 153, 252, 285
failure rate, 265
faulty realisation, 164
feature, 16, 146, 252, 285
formal parameter, 253
formation, 59, 247, 285
framework, 31, 121, 147, 268, 285
framework reference, 262

generalisation, 127, 285
generic model, 186, 194, 286
Ginganda, 117
Grampus, 286
graphical user interface, 7
ground weft, 44

guarantee, 11, 55, 153, 243, 286
guarantee:, 17, 125, 252
GUI, 7

HCI, 223
heuristics, 150
history, 121

Imbirfa, 117
Imkwezi River, 117
implementation domains, 230
implementation framework, 232
inadequate resourcing, 165
inertia problem, 166
information gathering, 286
Information Technology (IT), 286
inheritance, 129, 152
initiation, 132, 263, 275, 286
interact event, 19, 130, 264, 276, 286
interaction, 260, 273, 286
intrinsic, 27, 57, 163, 244, 286
invariant, 286
invariant:, 17, 251
Invitation to Tender, 210
involvement, 263
IT support, 177
IT system, 30, 103
ItemAllocator framework, 195
iteration, 150

jacquard, 43, 73
justification, 56

kit bag, 194

lifting box, 43, 73
local definition, 262
local reference, 262, 276
loom, 43

maintainability, 225
method, 286
model, 115
Model-Views System, 205
modelling language, 5, 115

monitoring and control problem, 167
mutation, 263, 275

narrow fabric loom, 43
New World, 28, 286
NIMWeC, 40
non-functional requirements, 153, 286
non-OO methods, 233
null behaviour, 261

object, 16, 286
object life history, 134
object orientation, 7
object state, 124
Old World, 28, 54, 286
OO design method, 232
operation, 125, 287
operator interface, 109
organisational structure, 119, 170
overlap, 144, 257, 270

parallel composition, 284
parameter, 131, 287
participant, 287
participation, 19, 263, 274, 287
patch panel, 44
pathology, 27, 89, 287
performance, 110, 223
permanent, 132
pick, 44
piece, 44
portability, 225
predicate, 254
prescription, 27, 89, 287
problem statement, 8
process, 83, 136, 287
process design, 233, 287
process support, 229
product development, 188
promotion, 30, 248, 287
prototyping, 222, 223
purpose, 6, 25
purposive entity, 116, 287
purposive mismatch, 164

purposive model, 16, 26

qualifier, 12, 56, 243

rationale, 229
reliability, 110
reliability requirements, 224
reliance, 11, 55, 153, 243, 287
rely:, 125, 252
repeat, 41
requirements specification, 8, 208
reuse, 231
ribbon, 44
rich picture, 9, 45, 160, 287
role, 11, 26, 55, 243, 287
Rumbabwe, 117

scope, 4
security requirements, 224
semantics, 186
sequential composition, 137, 284
service, 26, 71, 287
service provision, 11, 55
shed, 44
shuttle, 44
signature, 252
Soft Systems Methodology, 233
software properties, 225
specialisation, 127, 152, 287
standards and statutes, 225
state, 287
state transition model, 134
statechart, 134
static semantics, 288
static structure, 147, 288
statics, 16
status, 131, 287
stereotypical object, 19, 122
Stock Control System, 205
stop behaviour, 261
structural framework, 149
subclass, 288
subrole, 13
superclass, 288

syntax, 288
system, 3, 288
system boundary, 29

tailoring, 25, 157
taxonomy, 152
temporal framework, 149
termination, 132, 263, 275, 288
terms of reference, 46
timeline, 19, 77, 121, 274, 288
timing requirements, 224
trace, 91
transient, 132, 263, 275

validation, 210, 288
viewpoint, 6, 26, 288

warp, 44
weaving frame, 73
weft, 44